Ether
and
Reality

*A Series of Discourses on the Many
Functions of the Ether of Space*

Oliver Lodge

CAMBRIDGE
UNIVERSITY PRESS

CAMBRIDGE UNIVERSITY PRESS

Cambridge, New York, Melbourne, Madrid, Cape Town,
Singapore, São Paolo, Delhi, Mexico City

Published in the United States of America by Cambridge University Press, New York

www.cambridge.org
Information on this title: www.cambridge.org/9781108052665

© in this compilation Cambridge University Press 2012

This edition first published 1925
This digitally printed version 2012

ISBN 978-1-108-05266-5 Paperback

Cambridge University Press has long been a pioneer in the reissuing of out-of-print titles from its own backlist, producing digital reprints of books that are still sought after by scholars and students but could not be reprinted economically using traditional technology. The Cambridge Library Collection extends this activity to a wider range of books which are still of importance to researchers and professionals, either for the source material they contain, or as landmarks in the history of their academic discipline.

Drawing from the world-renowned collections in the Cambridge University Library and other partner libraries, and guided by the advice of experts in each subject area, Cambridge University Press is using state-of-the-art scanning machines in its own Printing House to capture the content of each book selected for inclusion. The files are processed to give a consistently clear, crisp image, and the books finished to the high quality standard for which the Press is recognised around the world. The latest print-on-demand technology ensures that the books will remain available indefinitely, and that orders for single or multiple copies can quickly be supplied.

The Cambridge Library Collection brings back to life books of enduring scholarly value (including out-of-copyright works originally issued by other publishers) across a wide range of disciplines in the humanities and social sciences and in science and technology.

Ether and Reality

ETHER & REALITY

A SERIES OF DISCOURSES ON THE MANY FUNCTIONS OF THE ETHER OF SPACE

BY

SIR OLIVER LODGE

F.R.S.

D.Sc. (Lond.); Hon. D.Sc. (Oxon.); Hon. Sc.D. (Cantab.); Hon. LL.D. (St. Andrews, Aberdeen, Glasgow, Edinburgh); Hon. D.Sc. (Liverpool, Manchester, Sheffield, Adelaide, Toronto); Hon. M.A. (Birmingham); Mem. Inst. Elec. Engineers; Hon. Mem. Lit. Phil. Soc. (Manch.); Corr. Mem. Amer. Phil. Soc. (Philadelphia); Corr. Mem. Accad. Sci. dell' Istituto (Bologna); Corr. Mem. Bataafsch Genoots (Rotterdam); Hon. Mem. Inst. Elec. Engineers; President of the Radio Society; Rumford Medallist of the Royal Society; Albert Medallist of the Royal Society of Arts, as Pioneer in Wireless Telegraphy; Ex-President of the Physical Society; Ex-President of the British Association; Ex-President of the Röntgen Society; Ex-Professor of Physics in the University College of Liverpool; Ex-Principal of the University of Birmingham.

HODDER AND STOUGHTON

LIMITED   LONDON

First Printed .   .   .   . 1925.

Printed in Great Britain by Richard Clay & Sons, Limited,
BUNGAY, SUFFOLK.

CAMBRIDGE LIBRARY COLLECTION

Books of enduring scholarly value

Physical Sciences

From ancient times, humans have tried to understand the workings of the
world around them. The roots of modern physical science go back to the
very earliest mechanical devices such as levers and rollers, the mixing of
paints and dyes, and the importance of the heavenly bodies in early religious
observance and navigation. The physical sciences as we know them today
began to emerge as independent academic subjects during the early modern
period, in the work of Newton and other 'natural philosophers', and numerous
sub-disciplines developed during the centuries that followed. This part of the
Cambridge Library Collection is devoted to landmark publications in this area
which will be of interest to historians of science concerned with individual
scientists, particular discoveries, and advances in scientific method, or with
the establishment and development of scientific institutions around the world.

Ether and Reality

Among the widely agreed facts of physics in the late nineteenth century
was the existence of luminiferous ether: the medium through which light
was thought to travel. Theorised to be a highly rarefied substance, the ether
accounted for the movement of light, gravity and even heat across a vacuum.
It also had great implications for spiritualism. Where thought was not proven
to be a result of chemistry in the brain, the presence of ether allowed for the
idea that cognition and emotion might exist independently of a physical
body. First published in 1925, this monograph by the eminent physicist
and ether advocate Sir Oliver Lodge (1851–1940) was written for the non-
scientific reader. With a focus on straightforward explanations rather than
mathematical theory, his book still represents a fascinating introduction to
the topic today.

REALITY is what everyone is keen to know about. No one wants to be deceived; all are eager for trustworthy information, if it be forthcoming, about both the material and the spiritual worlds, which together seem to constitute the Universe. The Ether of Space is the connecting link. In the material world it is the fundamental substantial reality. In the spiritual world the Realities of Existence are other and far higher; but still the Ether is made use of, in ways which at present we can only surmise.

An attempt to set forth in intelligible fashion something of what is known about The Ether and its Functions is made in this book: and although like everything human it is far from infallible, it represents the outcome of a lifetime of study and meditation, and may be acceptable as a guide until something still better is available. The spelling etherial or etheric is here used for the adjective, instead of the more poetic ethereal, because the ether is dealt with not as a rarefied essence but as a substance with ascertainable physical properties, to which the ideas usually and properly associated with the word "ethereal" are foreign. It has been impossible to treat all parts of the subject adequately in a small book mainly addressed to the non-expert, but this elementary treatment may prepare the way for a larger and more technical volume later in the year.

OLIVER LODGE.

11th March, 1925.

Contents

B

Prologue on Science and Philosophy

Yet all experience is an arch wherethro'
Gleams that untravell'd world, whose margin fades
For ever and for ever when I move.
                              TENNYSON, *Ulysses.*

WE live in a Universe of which we know
very little : we eke out our knowledge by
precarious reasoning. We are apt to con-
front a special instance with a generalization; as if we should try to establish the
fact—otherwise doubtless certain enough—
that, say, Œdipus and Moses and Elijah
must have died, by the thesis, All men are
mortal, and they were men. But admitting
the minor premiss, how do we know that
their end was not exceptional? The generalization sums up our experience, but *proves*
nothing,—*i.e.* adds nothing to our certainty.

An induction is helpful as a summary, but
has no power over a specific instance to the
contrary, if such instance were forthcoming.
Proof might be based on the known liability
of organic matter to degeneration : but

13

apply this to an amœba and it fails; a lowly organism may be killed, but it need not die. So we might argue thus :—cells may persist, a body is composed of cells, therefore it may persist. Experience disproves the conclusion, it is a *non sequitur*. Reasoning of a formal kind is based on generalizations which may or may not be true; whereas experience can be direct and specific as well as general.

Some things there are however which evade experience, and these are apt to be excluded altogether by the inductive process on which all generalizations are founded. It should be more widely recognised that inductions may be hasty, or may be extended unduly : a single instance, well established, will upset a premature generalization. The force of prejudice, however, may be so strong as to overpower the evidence of our senses, or render us incompetent to recognise their clear indications. As if an unwary naturalist, having unconsciously assumed as a major premiss that all grass is green, might thereafter blind himself to the direct experience that the blade of grass shown him is red. Similarly, many people would insist that communication with the dead is

14

impossible. " The dead know not anything. In that day their thoughts perish." A major premiss is usually a generalization from necessarily imperfect knowledge, an assumption which may be upset by a single authentic specific instance: it has or ought to have no force, save a cautionary one, against direct experience.

The aim of science is not mere formularies or convenient modes of expression, it is or should be truth and reality; we seek, on the evidence of our senses, to form some conception of what Reality is like. But some kinds of reality are evasive; they either never make an impression on our senses, or very seldom. What strongly appeals to us is matter—material objects in various forms; anything else seems, or really is, a matter of inference. The inference may strike us as of doubtful validity unless we can adduce some direct sensory experience in its favour; and this evidence it may be difficult to receive or attach credence to, in the face of a uniformly established prejudice which we may dignify as a law of nature with no exceptions. So it happens that even direct experience of an exceptional instance, combined with evidence or testimony concerning many such instances, is sometimes powerless to effect conviction. Indeed persons afflicted with what psycho-analysts of the more reasonable variety might call " a materialistic complex " * appear constitutionally unable to open their minds to evidence of any non-material or anti-materialistic kind. They straightway deny its authenticity. It makes no impression; they are incompetent to receive it, however cogent; though otherwise and in other directions they are mentally alert and highly qualified intellectually. They exhibit a curious kind of mental aberration or un-conscious warp, and yet they are quite unconscious that their perceptive faculty is atrophied in one direction. They have been known to stigmatise as demented those who call their attention to exceptional specific occurrences. But there is no dementia on either side, there is only a kind of obsession; and the obsession is a very natural one, born of a wide knowledge, a broad basis of ex-perience, throughout which such exceptional incidents have not occurred, or at least have not forced themselves upon attention.

* W. R. Bousfield, K.C., F.R.S., on *A Neglected Complex* (Kegan Paul).

We can admit that normal experience displays mind only in association with matter. It is therefore pardonable though illogical to assume that without matter mind cannot exist; in other words, that a physiological organism is an absolute necessity, not only for the display, but for the very existence, of memory, thought, and affection, notwithstanding that these attributes make no pretence of being themselves physiological things. Memory is not really a function of matter, though it does seem to have a physical basis or concomitant. Love need not necessarily be associated with protoplasm. Thought is not proven to be a secretion of the brain. We have no right to make that over-hasty generalization : we ought to keep our minds open and be guided by carefully scrutinised facts. This cannot be done without patience and study; and study needs an open mind. Violent assertions, whether positive or negative, are useless, or are only useful for focussing attention : we must be guided by observed facts, not by preconceptions or past experience alone, else would new knowledge be impossible.

Similarly if we make the assertion that all

bodies—bodies of every kind—must be composed of matter, we are speaking plausibly, but may be generalizing too hastily and leaving out of consideration the possibility that some "bodies" may be composed, either wholly or partially, of Ether. Such an idea would not have occurred to us unless the idea of Ether had been brought to our minds in a multitude of other ways, and unless we had begun to realise something of its reality and its numerous or perhaps innumerable functions. Hitherto it has been almost ignored in Philosophy. That is a defect to be remedied: it is no fault of philosophers. Physics must come to the rescue. Philosophy seeks to unify and comprehend all knowledge, and cannot afford to ignore anything—certainly not so omnipresent and intense a reality as the Ether of Space. For to know anything thoroughly, nothing accessible must be excluded. It is only because the Ether has seemed inaccessible that it has been neglected. Neglect was inevitable until more was known about it, and hitherto Physicists have not sought to expound what in this direction they know.

Many of us are not even physicists, but

just learners; such as both young and old ought to be. Young people must take things piecemeal, and be content at first to learn something about them. They cannot learn all about them, and they cannot philosophise; they do not know enough. Probably we none of us know enough to philosophise effectively. Even philosophers have to do their best with less than complete knowledge; and accordingly each makes a system, with which others do not agree. Each however probably catches a glimpse of some aspects of truth, and that is why the history of Philosophy is instructive.

We can all recognise the very certain truth that to know all about any one thing we have to know about a great number of other things. Everything is interlocked; we cannot take a comprehensive survey before we look at things individually, and we cannot consider individual things fully and completely without a comprehensive survey. Thus there is a difficulty, but it is unavoidable.

In science, as a rule, we concentrate on one aspect, and try to get that clear. Hitherto science has mainly concentrated on the purely material aspect of the Universe; while the philosopher is left to group all aspects together if he can. But there are gaps, which he must depend on Science to fill up. And sometimes he has to wait, not always knowing what he is waiting for; not always knowing that there is a great deal to wait for.

Meanwhile life is short, and if we do nothing but wait, no progress is made. Progress can be made, but always tentatively, and with a sense of incompleteness. Everything excluded is a weakness. To exclude the Ether is a terrible weakness: an effort to understand the connexion between mind and matter is hopeless if we exclude the *tertium quid*, the essential intermediary. To exclude life and mind is another weakness: it is the basis of a materialistic system. To exclude matter is another but less common error,—the basis of a narrow idealism. To over-emphasise conduct as a test for truth is the basis of Pragmatism. To under-estimate conduct and practical affairs is Mysticism. The positive side of all these systems may be strong; the negative side is feeble and misleading.

It is rather like the old controversy between faith and works. All sides must be represented in a complete scheme; and if by reason of frailty we cannot regard all aspects together, then we must regard them piecemeal or seriatim, and there must be division of labour. My business is to emphasise the Ether, and its bearing not only on matter but on life and mind.

To this end I am preparing a book of some size; but inasmuch as the subject continually touches on familiar problems— not those of physics and philosophy alone but those of humanity in general—it has been thought well to broadcast and publish separately this general summary of the argument.

I do this the more readily, and with a greater sense of responsibility, because I believe that our present physical knowledge, when properly grasped and accepted, constitutes a beneficent source of power, a fertilising influence, a body of illustration and parable, which can be drawn upon and used by those whose business it is to deal with still Higher Things. Their congregations may not know enough, they themselves may not know enough, to utilise ascertained facts to the full. If they did,—if only they could apprehend a tithe of what is now

known by specialists,—their teaching would
be suffused with a dominating sense of
Reality, in the strength of which they would
press forward with an energy and enthusiasm
such as were aforetime evoked in one to
whom a great experience had been vouchsafed, and who continued not unmindful of
the Heavenly Vision.

Chapter I   *The Ether and its Vibrations*

The vast interplanetary and interstellar regions will no longer be regarded as waste places in the universe. . . . We shall find them to be already full of this wonderful medium; so full that no human power can remove it from the smallest portion of space, or produce the slightest flaw in its infinite continuity. —CLERK MAXWELL.

WHAT fills empty space? What is there between the worlds? Not air: the atmosphere soon stops, and beyond there seems nothing—nothing appreciable, only intense cold.

" The wind that blows between the worlds, it cut
him like a knife,"

says Rudyard Kipling concerning one Tomlinson. Well, that is the Ether; it is absolutely cold. We on the comfortable earth are five hundred degrees Fahrenheit warmer. Five hundred degrees hotter would melt some metals : five hundred degrees colder is the temperature of Space. Space is full, not of matter, but of Ether. The Ether is other than matter; and it fills all space in the most thorough manner: there is nothing so omnipresent and so efficient in the physical universe.

We enter on some explanatory talks about the Ether of Space, describing its various functions or uses, emphasising its importance in the understanding of the Universe, and showing how essential it is to all Reality

We employ the Ether every day and every minute of our lives; it is the very breath of our material existence; but we usually think little about it. Some few even deny its existence. This is ungrateful : the result of a temporary uncertainty which can be removed.

It is commonly said that we have no sense organ for the appreciation of the Ether; and we have not any means of appreciating it directly, but we are very much accustomed to appreciate the phenomena which go on in it, or in other words to apprehend its modifications. It might almost equally well be said that we have no sense directly to appreciate the air we breathe; and I suppose that children have to be taught that an atmosphere really exists round the earth. We can however appreciate its modifications, as when there is a wind; or if we remove some air from a vessel, making a partial vacuum—say by throwing in a bit of burning paper—and close the orifice with our hand, we can feel the atmospheric pressure; which is very great, though so uniform that we usually ignore it, and it had to be discovered by Galileo, Torricelli, and Pascal. Apart from experiment however we can appreciate the vibrations of the air, for they are what we call noise or sound or music; but of course children of many ages could hear music without knowing that the atmosphere had anything to do with it : just as they can look at pictures without recognising the mediation of the ether.

Probably a wind is the most direct method of apprehending the air. Many attempts have been made to detect a wind in the Ether : the Earth is moving through it at a tremendous pace, and therefore, relatively, it must be streaming past us; but we can feel nothing of it : and, what is more, our most delicate instruments, specially designed to that end, can exhibit or detect nothing of it. Consequently it has been possible to doubt the existence of such a medium.

Again, we have no means of making a vacuum in the Ether and feeling its pressure :

we have reason to think that that pressure is enormous. The *air* pressure is a ton to each square foot : an ordinary barometer demonstrates this; it is an instrument for measuring the pressure of the atmosphere. But there is no instrument for measuring the pressure of the Ether, which is probably millions of times greater : it is altogether too uniform for direct apprehension. A deep-sea fish has probably no means of apprehending the existence of water, it is too uniformly immersed in it : and that is our condition in regard to the Ether.

But we can feel its vibrations. Hold your hand in front of a blazing fire ! It is not hot air that we feel; the air remains cool : it is not heated by radiation. What we feel is due to Ether vibrations : they excite the nerves of the skin, and give us the tingling sensation which we call heat or temperature. It is really the temperature of the skin that we feel, but it is excited by the tremors in the Ether. Again, if we sit in the sunshine we can realise, if we attend, that we feel all round us a quiver of the Ether : it may actually bronze our skins, and on the top of a mountain may raise blisters : it may produce an effect which

28

overpowers the automatic temperature-regulation of the body, and may thus cause what is called sunstroke. All these skin sensations are directly due to the Ether and its vibrations. The vibrations originated in the Sun, and have travelled 92 million miles of cold empty space, taking eight minutes on the journey before they reach us. They achieve many results on arrival :—Photographic chemicals are blackened by the vibrations. Leaves of trees make use of them : every green leaf elaborates crude sap by aid of solar energy and turns it into feeding material or vegetable tissue. All vegetation grows at the expense of the Ether tremors in which it is immersed : plants extract energy from the Ether and store it in their substance : store it, it may be, for hundreds or thousands of years; so that when you make a wood or a coal fire you liberate the stored or dormant energy, the ether recovers it and gets it back again in perceptible form.

Every part of the skin is sensitive to the Ether quiver; but a particular region is localised in nearly all animals so as to be especially sensitive, and is called an eye. Yet " seeing," though it tells us about objects, tells us nothing about a medium and its vibrations, nothing about the machinery and vibrations by means of which we see. It is quite possible to live amongst such vibrations and to know nothing about them : most people take the phenomenon for granted and do not analyse the cause. Only a few of the human race began to detect what was happening; and those few lived in the beginning of the Nineteenth century. Late in time the knowledge came. After that, it was either ignorance or else some opposition theory which enabled people to ignore or deny the Ether. But when, later on, we began to make experiments in electricity and magnetism and use them for ordinary purposes, the ignoration of the Ether was more difficult. We are using it now in radio telegraphy : but before that we used it to drive our street cars and machinery : in America they use it for judicial killing. It seems madness now to deny the existence of a thing which is all round us, which we daily utilise, and which is constantly exerting influence upon us.

Perhaps I may be allowed a parable, and say that similarly it is customary to ignore or to deny the existence of a spiritual world. As with the Ether, so we are painfully unaware of such an influence, we are blind and deaf to its reality : we are really stupid in our self-satisfaction and narrowness of outlook. Most of us apprehend little more than the animals : we go on with no more knowledge than is necessary for daily life : and it is rather the fashion to despise those who seek to enlighten us. There has always been a tendency to deny and contemn the pioneers of knowledge : all knowledge has to make its way slowly and painfully against a mass of prejudice and inertia. To a certain degree this may be wise : to accept things too readily and prematurely would be rash : unbalanced and enthusiastic credulity is always to be deprecated : it is better to be slow in accepting the truth than to be ready to accept falsehood. A certain amount of opposition is salutary; though there comes a point at which opposition to truth ceases to be beneficial and becomes mere obscurantism. Opinions may differ as to when that stage is reached in any given case.

Meanwhile our theories or our lack of theory, our perception or our blindness, do not alter facts. The facts are there all the time, and are independent of what humanity thinks about them. For ages people were ignorant that the stars were other suns, and not mere appendages to the Earth. We are constantly closing our minds to Reality, or rather we are living among realities of which we have no apprehension. Open our eyes that we may see! Should we not ask this? Some things in this age of science we are beginning to learn which were unknown to the Ancients; but in time we too shall be Ancients, and our descendants will wonder at our blindness and stupidity. Especially perhaps at the blindness and stupidity of some of the learned—some of the scribes.

Revelation has not ceased, though it takes many forms; and those to whom perception has come must relieve their minds by utterance, and not expect to be fully understood in their day and generation. Their own knowledge, so far as it is sound, is but a step in advance; and they must have faith that humanity will take it in due time. Meanwhile they must possess their souls in patience; they need not strive nor cry; in quietness and confidence shall be their strength.

Chapter II   Fundamental Notions about an Ether

E

Chapter II Fundamental Notions about an Ether

How matter is held together, and how we see it.

Apollonius of Tyana is said to have asked the Brahmins of what they supposed the Cosmos to be composed.

" Of the five elements."

" How can there be a fifth," demanded Apollonius, " beside water and air and earth and fire? "

" There is the ether," replied the Brahmin, " which we must regard as the element of which the gods are made; for just as all mortal creatures inhale the air, so do immortal and divine natures inhale the ether."

THE first thing to realise about the Ether is its absolute continuity. Let me explain. Matter is discontinuous; it consists of portions with gaps between. You see this clearly enough in the stars; they are bodies separated by wide empty spaces, they are not massed together. There must be a reason for this; the reason is partly known, but is not easy : we will be satisfied with the fact that it is so. Matter is full of discontinuity; it consists mostly of empty space : the portions of matter in space are all well separated from each other in proportion to their size. Fire an infinitely long-range projectile into the sky, and the chances are it will not hit anything : Lord Kelvin reckoned that the chance of hitting anything by such a projectile was about the same as the chance of hitting a bird if you fired a gun at random. That is one of the first things to realise about matter : there are great gaps between its particles.

You may say, That is all very well for the sky and the stars and planets; but what about the earth ? What about a piece of rock, or furniture, or any solid object ? Do you mean to say that the particles of a body like that are widely separated, with great spaces between them in proportion to their size, and that a straight line might penetrate them deeply without encountering a particle? Yes, I do : that is what I mean by the discontinuity of matter. It is discontinuous on a small scale as well as on a large scale. It does not appear so, but that is only because our senses are not fine enough to tell us about things on a small scale : we can only see things on a big scale. A microscope is of some assistance, but nothing like sufficient : no microscope, however powerful, can show us an atom, still less can it show us how an atom is composed and how far apart its ultimate particles are : we know this otherwise and indirectly. It is however common knowledge, now, that matter is built up of minute electric charges, both negative and positive, which are called electrons and protons. It is also known that these electric units are so extremely minute that they are separated from one another like the planets in the solar system : the greater part of the atom is empty space, just like the sky on a small scale. Or, more clearly, if we could take a solid body and magnify it sufficiently (which is impossible) we should see it something like the night sky.

Since the particles of matter are thus separated from each other and never in contact, it would seem to follow that they are all independent of each other, disconnected, nothing uniting them,—the particles completely separated by empty space. If there were nothing existent but matter, that would be so; there would be no unification, no binding force, no family relationship, nothing but separate independent particles : that is what would happen if Space

were really empty; and the universe would
be not a cosmos but a chaos.

We know better than that; we know
that the stars are not independent of each
other; they are bound together into systems :
there is a unifying and connecting force
between them, which is called Gravitation,
though it is not fully understood. Hence the
space between them cannot be really empty;
the interspaces must be filled up somehow :
there must be something which is without
gaps, something really continuous, something which combines the whole together,
welding all the separate bodies into a cosmos.

The same thing is true inside any solid
body : the separated particles cohere, they
are not independent of each other; there is
no chaos to be found anywhere. The solid
has a definite size and shape; and if it is a
crystal its shape may be beautiful and very
definite. There is evidently law and order
reigning among the particles : however great
the interstices between them, they must be
full of something : space is not really empty,
though it is empty of matter. Matter exists
as separate particles, here one, there another.
But the uniting " something " is not composed of particles at all; it is continuous :

38

it unites the particles with a force which is known as Cohesion.

What you choose to call this unifying " something " is of no consequence. The Ancients sometimes spoke of the " Ether," possibly as an addition to the usual four elements, and Sir Isaac Newton adopted this term for the connecting medium. The optical medium connects the particles together in a solid or a liquid, and the same medium connects the Heavenly Bodies together into systems and clusters and constellations and nebulæ and Milky Way.

All pieces of matter and all particles are connected together by the Ether and by nothing else. In it they move freely, and of it they may be composed. We must study the kind of connexion between matter and Ether.

The particles embedded in the Ether are not independent of it, they are closely connected with it, it is probable that they are formed out of it : they are not like grains of sand suspended in water, they seem more like minute crystals formed in a mother liquor. The mode of connexion between the particles and the Ether is not known; it is earnestly being sought : but the fact that there is a connexion has been known a long time. We know it, because a particle cannot quiver or move without disturbing the medium in which it is. A boat cannot oscillate on the surface of water without sending off waves or ripples; a bell cannot vibrate in air without sending out waves of sound; a particle cannot vibrate in Ether without sending out waves akin to those of light.

So the second thing to learn about the Ether is its property of conveying light. It seems curious to call it a " second " property, because historically it was the first,— the first discovered and the first on which attempts were made at elaboration. The Physics of the early part of the Nineteenth Century was almost wholly occupied with it : the highest genius was devoted to the theory of Ether waves, and the climax was reached by Clerk Maxwell. The whole of the immense Science of Optics is involved and grew out of it; but as with everything else it is difficult completely to understand and to realise clearly what is happening : certain things can be stated with apparent simplicity, but the full explanation is not yet attained.

The first and most definite fact, on which

there is complete agreement, is the rate at which Ether waves travel, the thing ordinarily called " the velocity of light." This speed is the most fundamental and absolute thing in the physical universe, and it is evidently related to some fundamental or constitutional velocity, the full meaning of which has still to be discovered. Meanwhile we can make elementary statements about what has been observed in connexion with it.

The speed is measured by timing the interval required by light to travel a certain measured distance, whether it be a distance measured on the earth or a greater distance measured in the heavens. The results all agree, and there is no doubt that all Ether waves, however else they differ, travel at the same pace. The speed of light is not only the speed of that by which we see things, but it is the speed with which every disturbance travels in the Ether of Space. Such disturbances may be the great waves (akin as it were to Atlantic rollers) which we employ in radio telegraphy; or they may be the small ripples which, when they break upon the shore of matter, excite heat; or they may be the minuter tremors which in enormous numbers enter the eye and operate the curious receiving mechanism there, so as to disturb the nerves and give us the sense of sight, or, by rearranging the chemicals on a glass plate or film, can reproduce the likeness of the objects which have emitted them; or they may be the still minuter tremors—small almost beyond imagination, and beyond the power of any microscope to utilise,—fearfully rapid tremors or ether vibrations which can be excited electrically in a form which we know as X-rays. But whether big or small, they all travel at the same pace, with a speed far beyond anything in our experience, a speed which it seems impossible even for the Ether to over-top.

Imagine a thread wrapped round the Equator of the world, crossing all the continents and oceans; stretch such a thread out into a straight line—that is the distance which light can travel in the seventh part of a second. To get the distance traversed by light in a second, the thread would have to be wrapped round the world seven times and then stretched out; such a thread would reach nearly to the moon. The light of the moon takes a second and a quarter to reach the earth : from the Sun it takes eight minutes : from the stars, even the bright stars, it takes years or even centuries : while some of the dim and distant objects revealed in a large telescope we see only as they were a hundred-thousand years ago. So immense is the scale of the Universe !

All this is well and even popularly known : the difficulties do not lie here, they lie in determining the exact nature of the waves and the way in which they are produced. We have to work by analogies for the most part. As a vibrating bell or string or tuning-fork excites waves in the air, so a vibrating electron excites waves in the Ether : the processes are analogous, not identical; and if we tried now to enter into more detail we should get beyond our depth.

Meanwhile if ripples are travelling from distant objects, there must be something which is rippling. You cannot imagine *Space* being thrown into Vibration; there must be something in space which vibrates, and that " something " extends to the furthest visible object. This was our first idea of the Ether of Space : it is more than a century old, and the argument was as valid in 1825 as it is to-day. The Ether was therefore called " the luminiferous

43

Ether," the light-carrier, the vehicle of light. Nor of light only, but of every other link between the worlds and between the atoms; the vehicle of Gravitation, as Sir Isaac Newton suspected; the vehicle of Cohesion too, as we now know; the unifying and connecting mechanism which welds together the disconnected atoms of matter and makes cosmos out of chaos.

However evasive the ether is to our senses it is a great reality and we continually use it. The waves emitted by a radio-aerial will get to the Antipodes, say New Zealand, in the 1/14th part of a second. How far will sound waves travel in the same time? Sound in air takes five seconds to go a mile. Consequently in the 1/14th part of a second they go the 1/70th part of a mile, which is 25 yards, that is to the back of a hall in which one might be speaking. Ether waves travel just about a million times as quickly as sound waves: consequently if the waves were of the same length, the vibrations would be a million times as rapid. But the Ether waves by which we see are not of the same length: sound waves are a few feet long: whereas a series of 50,000 light waves is only an

inch long. Consequently the rate of vibration which the eye perceives is 500 million-
million per second—a quite incredible number ! But in dealing with the Universe we
must not be afraid of large numbers : the
magnitudes we deal with are many of them
appalling, some of them appalling for size,
others for smallness, some for rapidity, others
for unknown and mysterious properties. We
have as yet very little acquaintance with the
Universe; sometimes we seem to know a
great deal, at other times we realise that
we hardly know anything. The mystery of
it all escapes us, the possibilities of it are
beyond our conception : many of them we
could not apprehend if they were explained
to us, we have not the terms or ideas to
understand them. Meanwhile we grope along
as best we can, and do our daily work and
have a keen expectation of the future; and
he is wisest who denies least of the mystery
which surrounds us and the possibilities
ahead. To assert requires knowledge, to
deny requires much more knowledge. Let
us be satisfied with positive knowledge, so
far as it has been vouchsafed to us, and
leave negations to the self-sufficing and the
omniscient. We can deny the self-contradic-

tory and the absurd, but in the Unknown and the Mysterious, denials have no legitimate place : our business is carefully and cautiously to ascertain what is. We are surrounded by infinity, infinities of various kinds; and the wealth of existence is such as to justify a Faith in our highest conceptions, a Hope in the possibilities which lie before us, and a Charity which enables us to do our daily work and to love our fellow men.

But that one ripple on the boundless deep
Feels that the deep is boundless, and itself
For ever changing form, but evermore
One with the boundless motion of the deep.
           TENNYSON, *The Ancient Sage.*

AN immense amount is known about waves, and the mere fact that a medium is able to transmit waves tells us something about its properties. A wave is always the result of an oscillation or vibration; and for an oscillation to be possible two things are necessary,—those two things are the power of recoil and the power of over-shooting the mark. The power of recoil is called elasticity; the power of overshooting the mark is called inertia. A bent or coiled-up spring possesses one: any kind of a load or mass possesses the other. Their conjunction may be most easily illustrated by a piece of elastic or a spiral spring, held at the top by a hand or a fixed support, and loaded at the bottom by a weight. When everything is quiet the weight is in a position of equilibrium; and if it is either pulled down or raised up a little and let go, it will dance up and down. Let us analyse that motion. It tends to return to its original position when let go, by reason of the recoil or elasticity of the spring, but it will not simply return and stay there; it will overshoot that position, and only recover it after several oscillations. The recoil or recovery is due to elasticity, the overshooting is due to inertia. If the load is a massive one, the oscillations are slow : if the spring is a stiff one, the oscillations are quick. The rate of oscillation depends simply and entirely on the ratio of the elasticity to the inertia.

If a string is tightly stretched and then plucked, the stretching of the string confers elasticity on it, and makes it recoil back to its old position; but the string has a certain momentum, and accordingly it overshoots that position and vibrates, giving a musical note if the vibrations are rapid enough. Stretch the string tighter, the note rises in pitch : elasticity is increased. Load the string, or otherwise make it more massive, the rapidity of vibration or pitch is lowered. It will be found that if the tension of the

On Waves

string is increased four-fold, the pitch rises
an octave; that is the rate of vibration is
doubled. Conversely, if the load is increased four-fold, the pitch falls an octave;
the rate of vibration is halved. This is
expressed by saying that the rate of vibration depends on the square root of the ratio
of elasticity to inertia. In a continuous
medium the inertia of any given volume is
called its density : it is also sometimes
called specific gravity. Mercury is thirteen
times as dense as water : water is 800 times
as dense as ordinary air : lead is denser,
that is heavier bulk for bulk, than iron :
iron is denser than aluminium : the densest
substances known are platinum and gold,
twenty times as heavy or massive as an
equal bulk of water. The lightest gas known
is hydrogen; air is about fourteen times as
dense as hydrogen. All these things are
familiar; but I remind you of them, because
they are one of the causes that regulate the
rate of propagation of waves through these
different substances.

The other property is elasticity : and in a
gas that is equal to the pressure, it measures
the incompressibility of the substance. The
incompressibility of water is much greater

than that of air, thousands of times greater : this more than compensates for the extra density, and accordingly sound waves travel in water four times as quickly as in air. But in the Ether, waves travel a million times as quickly : hence the ratio of elasticity to density in the Ether must be enormous. The important thing at present is to realise that the fact that the Ether can transmit waves tells us something about its properties, tells us that it must possess something akin to elasticity, giving it a power of recoil, and something akin to density or inertia, giving it momentum. We must not suppose that these properties are due to the same cause as those of matter, but they must have some analogy. The possession of those properties makes the Ether very real : and the fact that it can transmit waves at a definite and ascertained speed tells us already a good deal about it, and shows that it is a real substance, whose properties we mainly study under the heads Electricity and Magnetism.

Varieties of Elasticity

Elasticity may be possessed by a substance for various reasons. Gases and liquids shew it by their resistance to compression, *i.e.* their incompressibility,—a term which does not mean that they are incompressible, but that their incompressibility can be measured and specified ; the incompressibility can have a figure attached to it, can be expressed numerically in terms of proper units. A solid has other kinds of elasticity; it can be bent, it can be twisted; and in both cases it will recoil or recover from the strain, unless it is inelastic, like clay or putty, which are not able to transmit waves.

Waves then are of various kinds : sound waves are the simplest. The best known waves are those on the surface of the sea. The surface of water is naturally level; but if you lay a board on the top of a lake, say a wooden disk a foot in diameter, and then either lift it or push it down and let it go, it will oscillate : waves will be generated, and will spread out from the oscillating centre of disturbance, in accordance with its rate of oscillation. However the surface of water is deformed, it will tend to recover : it therefore possesses the power of recoil, a power which in this case is due to gravity. It also possesses inertia, so that a hump let go becomes a depression, and then a smaller hump again, and then another depression, and so on alternately, until the disturbance is wiped out by the waves it has generated, or by friction, or both.

A great deal is known about water waves, but that must suffice for the present purpose; they are only mentioned to show that elasticity may have various causes : the only thing essential is power of recoil or recovery. The recovery of a bent spring is one thing; it is this which drives our watches, and which gives the musical note in a harmonium or concertina (where the spring is called a reed); and this elasticity may be called its stiffness. The recovery of a stretched harp or violin string is another thing; it is due to the stretching force, and is commonly called its tension. The recovery of a raised weight is another thing : it is this which drives many of our clocks, and it is this which enables waves to travel along the sea. The recovery of compressed air or gases is another thing; it can be used to drive engines; it is this which enables waves of sound to travel.

The recovery of the Ether is another thing, and we do not *fully* know to what it is due. All we know about it is that it is not mechanical like the other things : Clerk Maxwell taught us that it was electrical, that it is the same property as we use when we charge a Leyden jar. We have no means of getting hold of the Ether mechanically : we cannot grip it or move it in the ordinary way : we can only get at it electrically. We are straining the Ether when we charge a body with electricity; it tries to recover, it has the power of recoil : if we charge it enough, it will, so to speak, snap its moorings, smash the insulating air, and give a spark. But like all the other things, it will not settle down into equilibrium instantaneously : the rushing electric current has momentum and overshoots the mark : so the body which had been positive becomes negative, and then positive again,—after the same fashion as the surface of water rises and falls alternately for a time; or as a thin steel bar clamped by one end in a vice, if the other end is pulled to the right and then let go, will oscillate over to the left and back again, several times.

That is how we start the wireless waves, or at least that is the simplest way. You charge an aerial till it overflows or sparks to earth; oscillations instantly begin; the aerial charges itself negatively and then positively again. The power of recoil depends on the properties of electric charge: the elasticity of the Ether is thus displayed. It is not matter which is strained when you charge a body electrically, it is Ether. It is really the Ether again which is strained when you bend a spring. Particles of matter are only moved or changed in relative position: it is always the connecting or cementing substance which is strained.

But what about the second property? Elasticity alone is not sufficient, an electric discharge must have momentum also: it must be able to overshoot the mark, else it cannot oscillate. In all the other cases we have been speaking of, this second power is due to inertia, massiveness, the fundamental property of matter. Ether is not any form of matter, as we know it: it seems strange that it should have an analogous property, a power of persistence, a power of obeying the First Law of Motion, that is of continuing to move until it is stopped, the power of overshooting the mark and going on even against an opposition force until that force is able to check it and throw it back again. Whether strange or not, the

Ether does possess this property, a property quite distinct from that of electric charge, a property distinct from anything that can be called elasticity, a property analogous to density or inertia. This property was studied long before its function in connexion with waves was known, that is long before the time of Clerk Maxwell; and it is commonly spoken of as Magnetism,—magnetism is the fundamental property of an electric current as distinct from an electric charge.

The Ether has two properties, an electric property and a magnetic property, the one corresponding to elasticity and the other to density. What either of them is due to, we do not fully know, we are trying to ascertain; but there they are, those two properties, and it is owing to their interaction that Ether waves exist : without them there would be no light, and of course there would be no radio telegraphy You can excite one without the other, but for waves you must have both. A charged body emits no waves, a magnet emits no waves; but you can only discharge a body by means of a current, that is to say, the act of discharge is a current. Magnetism is then evoked; the second property makes itself manifest,

H   57

oscillations occur, and waves spread out in all directions.

The pace at which waves travel depends on the ratio of elasticity to density, that is on the ratio of the electric property to the magnetic property of the Ether. We cannot specify these properties more completely in words : we have to use symbols,— symbols enable us to work out things we do not fully understand. When we fully understand a thing, we can express it in words : that time is not yet; and that constitutes one of the difficulties of the subject.

But it is really a difficulty which surrounds all subjects : we know nothing completely, and even our words are only symbols, though they are symbols we have got accustomed to; it may be doubted if they are really simpler or more expressive than algebraic symbols. Language is inadequate for dealing with the processes of nature in a complete and entirely satisfactory manner. We understand a certain amount, and try to express that; but we are aware of a mass of ignorance beyond. We know enough for practical purposes about matter and elasticity and inertia and other properties, but we do not know these things completely :

so also we know enough for practical purposes about electricity and magnetism, but we do not know them completely : we realise that there is much more to be known. So it is about everything really, though in some cases to all appearance our knowledge is greater than our ignorance : the truth of that however is very doubtful.

We have grown accustomed to the properties of matter, and to gravitation and the like, and so we delude ourselves into thinking that we understand them : we really understand less about them than we do about electricity and magnetism; these properties of the Ether are more fundamental than any of the properties of matter. We used to try to explain the properties of the ether in terms of matter : Lord Kelvin made a life-long effort in that direction, and at the end of his life he announced failure. The failure was due to no defect of his; such an explanation is impossible, Ether is not to be explained in terms of matter. We have learnt that the problem lies in the opposite direction; and modern Physics, since Lord Kelvin's time, may be said to be devoting itself to the explanation of matter in terms of Ether.

The two phenomena with which the human race has only comparatively recently become acquainted, the phenomena of electricity and of magnetism, are the fundamental things. Matter itself is an electrical phenomenon, just as light is : not electrical only, but magnetic also,—electromagnetic, as it is called. Light is electromagnetic; so is matter. Gravitation is an etherial phenomenon, we cannot say definitely that it is electromagnetic.

Those two properties of the Ether are known, but it may have others,—must have others, one would think. Gravitation may be a finger-post to some of the others : cohesion however is probably electromagnetic; and nearly all of the properties of matter will ultimately be expressible in those terms. Great strides already have been made in that direction : they seem to lead into an impenetrable thicket : we are not yet out in the open, it is hard going, and there is much to clear away. We had thought that the way towards the light must lie in the open country of ordinary mechanics; we are now plunging into the wood to seek a new way, changing our direction altogether; but flashes of illumination have been caught through the branches, which have heartened the younger generations of Physicists with a great enthusiasm. Obstacles are plenty, but they will not be deterred; they have caught glimpses of the light beyond : it may be long before they have made a path by which others can follow, it may be some time before they make a path for themselves : some may be lost in the under-wood, and there will be confusion and controversy for a time. But those who have insight and intuition know that here, through this strangely unpromising country, lies the road to Reality.

Chapter IV   The Ether as Transmitter of Force

Chapter IV   The Ether as Transmitter of Force
And what " Contact " is like.

For my own part, considering the relation of a
vacuum to the magnetic force, and the general
character of magnetic phenomena external to the
magnet, I am much more inclined to the notion that
in the transmission of the force there is such an action,
external to the magnet, than that the effects are merely
attraction and repulsion at a distance.   Such an action
may be a function of the aether; for it is not unlikely
that, if there be an aether, it should have other uses
than simply the conveyance of radiation.—Faraday.

Our commonest experience on this planet
is to see one body act on another. You see
a horse pulling a cart, or a magnet pulling
a bit of iron, or the earth pulling down
an apple : you also see a golf-club or the
head or the boot of a football player acting
on a ball, or a gun acting on a bird : or you
find yourself lifting a hat or opening a
window or throwing a stone. In fact in
the material universe you experience little
else than the action of one body on another.
But there is an obvious apparent difference
between the instances; in one set there

appears to be contact between the two bodies; in the other set they appear to act on each other at a distance. And action at a distance seems to require more explanation than when there is contact. You may think of some other familiar but instructive instances of action at a distance. You can act on a distant dog by whistling to it, or by throwing a stick into water for the dog to fetch out; you can act on a distant person by either shouting or writing him a letter or telegraphing to him. But in all such cases you know, or ought to know, that there must be a medium, connecting you and the distant person, through which the action is transmitted : and it may occur to you to ask whether action is ever direct, whether there is always something akin to a projectile or to a connecting string, whether in fact a medium of communication is always involved.

This was a question which puzzled Sir Isaac Newton, and therefore may well puzzle us. The earth not only acts on stones and apples, it acts on the moon, across apparently empty space, certainly across space empty of matter. The sun similarly controls all the planets, and must be acting even on the most distant stars—although no effect is in that last case perceptible. Those who have contemplated closely all these actions have been led to ask themselves whether the apparent difference between them is an appearance or a reality, whether bodies *always* act on each other through a medium, and whether there is ever genuine contact. The result of cogitation has been to promulgate two opposition theses : one to the effect that *a body cannot act where it is not :* the other that *true contact does not exist.* In other words, that in cases of apparent contact there is really an interval, though an extremely minute one, separating the two bodies, across which infinitesimal interval the force has somehow to be transmitted.

At first sight this denial of real contact seems an unnecessary complication, reducing the more obvious cases to the more mysterious kind. For contact seems intelligible, whereas action at a distance seems to demand explanation. Let us consider the two theses further.

First " a body cannot act where it is not." That sounds in accordance with common sense. But then the question arises :

Where is it? Is the magnet limited to the bit of steel that you handle and see, which is plainly limited or localized in space, or does it carry with it a field of force, which extends outwards in an unlimited manner, though with rapidly diminishing intensity? This field of force would then be essentially the magnet, though an intangible and supersensuous part of it. No one can deny that a magnet acts on a distant piece of iron; but it is quite possible to say that this only shows that the magnet extends as far as the iron, that its field of force reaches the iron, and that it is that extension of the magnet that produces the effect. This idea would answer the question, whether a body can act where it is not, in the negative; but would add that in some sense the body must extend beyond its obvious boundary, so that it really reaches everything on which it can act.

So Newton endowed the earth and the sun and planets with gravitational fields of force, which were an essential though not an obvious part of their constitution; and it becomes merely a question of convenience whether we say that the field of force of the earth reaches to the moon, or whether

we say that the earth acts on some intervening medium, and that this medium transmits the action to the moon. The facts are
clear and certain enough. The action is *as
if* the earth attracted the moon. All can
agree to that; though Newton felt, and
many others feel, bound to look for some
mechanism or medium of communication
which renders such attraction—such action
at a distance—possible. It is only the best
mode of expression which is doubtful; and
we realize that in order to express the action
completely we must know more about the
nature of gravitation.

Now take the other thesis, that " contact
does not exist." At first sight that looks
like nonsense. You see a book resting on a
table, you see a locomotive propelling a
truck, or a nurse pushing a perambulator :
contact between material bodies is obvious.
But when you contemplate the matter closely
and realize that matter is composed of
atoms, and that each atom is composed of
electric charges, you perceive that what you
call contact is not so simple. If in an
ordinary solid you have to admit that the
ultimate particles of which it is composed
are separated from each other in the same

sort of fashion, and by much the same relative distances, as the planets in the solar system are separated from each other—*i.e.* with intervening spaces great in proportion to their size—you naturally begin to doubt whether two different bodies can ever be more closely in touch than are the particles of the same body. And even in cases of impact, as when a hammer strikes a nail, or a ball rebounds, or two billiard balls kiss, you begin to wonder (if you are philosophically inclined) what the contact, atomically considered, is really like—whether it is real or only an appearance.

Astronomy gives us a hint in this connexion. You know perhaps that a comet coming out of the depths of space approaches the sun with constantly increasing speed, getting quicker and quicker as well as brighter as it comes nearer, that it then as it were rebounds from the sun and travels back into space whence it came, having acquired a conspicuous tail as the result of the collision. And if we were looking at the comet from a great distance, without being able to analyse exactly what was happening, we could very well liken it to the bounce of an indiarubber ball : the comet would appear to have struck the sun and rebounded. But we, being astronomically " near " the scene of action, are able to analyse this large-scale phenomenon more closely; we can tell (or astronomers can) that the comet has never struck the body of the sun, it has all the time been linked to it with gravitational hands across, as in a sort of Ladies' Chain, till it has been swirled round in a violently curved path and so been flung away again, travelling back by reason of its own momentum.

Here then is a collision which we can analyse : and we naturally ask whether an atomic collision (which is on so minute a scale that we cannot so easily examine it) is after all of the same nature. We find, or Sir Ernest Rutherford finds, that it is of just the same nature. Electrified particles, such as constitute the atoms of matter, join hands across, not of a gravitational but of an electrical kind, and swing each other round in exactly that cometary fashion, if their charges are of opposite sign. If their charges are of similar sign, so that they repel each other, the two particles do not approach so close : they rebound at a greater distance. They still do not strike each other; there

71

is, as it were, an elastic cushion between them; but the ultimate effect is precisely the same. An object or a particle approaching a repulsive centre slows down instead of hurrying up, storing up elastic energy in the medium between them, which then drives it back in a symmetrical path in the general direction whence it came. It takes the same identical path, seen from a distance, as if it had been attracted instead of repelled.

There are thus among electrified particles two different kinds of collision, the attractive and the repulsive kind; and these produce just the same effect, though for opposite reasons. No one seeing the result could tell which kind of collision had happened, whether it was of the attracting or the repelling variety. In neither case has there been any contact between the bodies themselves; the forces acting may have been violent, but they have been forces between the fields of the two bodies, or, in other words, there has been a strong deflecting force exerted in the medium between them.

A collision and rebound of this kind is an etherial phenomenon, conducted according to the law of the inverse square, just as in the case of a comet. There is no contact between the material particles themselves; in that sense it may be truly said that contact does not exist. There is always an elastic cushion which keeps the bodies apart : they both modify the ether in their neighbourhood; and a book resting on a table is really reposing on a cushion of ether.

Two pieces of glass or of metal pressed together do not really touch; if they did they would cohere; they remain separated by a microscopic distance. If they are pressed very tightly together and are very clean, or if a film of cement is placed between them, they may adhere and become one piece. Even then they are not really in contact, but they are closer now than the microscopic interval represented by a crack; they are now as close together as are the particles of any single body, their distance apart is molecular and ultra-microscopic. But even then they are cohering by their fields of force : the ultimate particles do not touch. Cohesion is an etherial force, pretty certainly either electric or magnetic, or both : the particles now no longer rebound; they have, as it were, entered into combination and attained a position of equilibrium.

But could it ever happen that two particles really collided? It could not happen if they were similarly charged, unless one was moving at a tremendous pace; they would usually repel each other so violently that they must rebound before they touched. But if they were oppositely charged, so as to attract, a real collision might occasionally occur; it would be very rare under ordinary circumstances, and something striking would be bound to happen. They might, as it were, obliterate each other, and we should have to ask what has become of their energy. They could not really knock each other into non-existence : depend upon it that never happens; energy is never destroyed. Some effect would remain; and it so happens that we know what that effect would be. A wave or pulse of energy would be generated in the ether : a luminous or X-ray pulse would be emitted by the collision, and would immediately start out in all directions with the speed of light. It is believed that in some of the stars such real collisions actually occur, and that this accounts for their violent radiation, which is of an intensity far greater than anything we experience on the earth.

Every collision or sudden change of speed generates some radiation. The molecular collisions with which we are familiar in a fire are of a gentle character : no destruction of matter occurs in them, at least no destruction of the permanent kind of matter which our instruments enable us to weigh and measure : only the extra energy of their motion is transformed into radiation. Moving matter has an adventitious or extra mass associated with it, and that kind of temporary mass can be shattered off into radiation by the ordinary encounter between their fields of force. Indeed, that is precisely how all radiation is generated.

In a wireless aerial the electrons are rushing along and being reflected. They are thus put into a state of large-scale oscillation, and the waves emitted are comparable to the dimensions of the aerial. Small waves, such as we call " light," are generated by oscillations of molecular dimensions; and X-rays are generated by oscillations or collisions of atomic dimensions, or sometimes even of the still smaller dimensions of an electron. There is no other way of generating radiation : it is all generated by sudden changes in the speed of particles, and is an outcome of their close relation or interaction with the ether.   The ether, as it were, stands by, always ready to pick up any loose energy and broadcast it with the speed of light.

Chapter V   Action at a Distance

That one body may act upon another at a distance, through a vacuum, without the mediation of anything else by and through which their action may be conveyed from one to another, is to me so great an absurdity that I believe no man, who has in philosophical matters a competent faculty of thinking, can ever fall into it.—NEWTON.

WITH the information of Chapter IV in mind, let us return to the two opposition theses. "A body cannot act where it is not." "Contact does not exist." We now see that there is a sense in which they are both true : we may have occasion to hint that there is another sense in which they are both false. But first let us see in what sense they are both true.

A body cannot act where its influence is not; if we put it in that form it becomes obvious. The whole point then turns on where the influence of the body is and what the influence consists of. If we identify the influence with its field of force, we can say that that field travels about with the body and forms part of it, usually an imperceptible and intangible part. Another and perhaps a better way of putting it is to say that one body can only act on another through a medium, a medium of communication. When a horse pulls a cart it is connected by traces; when the earth pulls the moon it is connected by the ether; when a magnet pulls a bit of iron it is connected by its magnetic field, which is also in the ether; and the same is true when the parts of a body cohere. Always look for the medium of communication : it may be an invisible thread, as in a conjuring trick; it may be the atmosphere, as when you whistle for a dog; it may be the ether, as when you beckon to a friend; or it may be a projectile, as when you shoot an enemy. Or, again, it may be ether ripples, as when you look at a star.

You cannot act at a distance without some means of communication; and yet you can certainly act where you are not, as when by a letter or telegram you bring a friend home from the Antipodes. A railway signalman can stop a train or bring about a collision without ever touching a locomotive. A Board of Directors sitting in London or Edinburgh can permanently span the Forth. A conclave of German politicians could, and did, operate on innumerable families in England and slaughter their most promising members without the direct action of a finger. A breed of mosquitoes for a long time rendered impotent the design of a Panama Canal, without malice and in the ordinary course of their existence, until Sir Ronald Ross and others with their microscopes made the enterprise possible. The interlocking of events is so complicated that actions may spread far beyond the apparent scope of an individual, and entail consequences unintended and unthought of. We may well feel inclined to ask : Is mental action ever anything but " distant " ? Does the mind always act on matter through some intermediary, such as an etheric medium ?

Now turn to the other assertion : " Contact does not exist." More accurately, Material particles never come into contact; they are cushioned from each other by attractive or repelling forces. They act on each other through the ether, just as essentially as when a magnet acts on a bit of iron, or as the sun acts on a planet. Cohesion does not mean direct contact, but residual electrical attraction or molecular affinity

L   81

across minute intervening space. The particles of matter are all discontinuous, but
are embedded in a continuous ether; and
when we say that contact does not exist, we
only mean contact of matter with matter.

Hold out your hand, and you feel the fire.
What do you feel? Not the fire direct, but
the ether ripples excited by it. Something
is in contact with your hand, but it is not
anything material : the ether connects your
hand and the fire. The fire acts on your hand
from a distance, and we know how the action
takes place. The earth acts on the moon
from a distance, and we do not fully know how
the action takes place. Attraction is an
" as if " mode of expression. Einstein attempts to get behind it and replace action
at a distance by a contact effect. A strain
in the ether we should wish to say. Einstein
does not say as much as that, but he recog-
nizes that it must be something directly in
contact with the moon that is curving its
path, even if it be only a warp in space. We
can adopt the expression and say that a large
body like the earth warps the ether all round
it, thus making other bodies fall towards
it as if they felt its attraction.

So when we take the connecting medium

into account, we find nothing but contact—not what is ordinarily called contact, but immediate connexion; there is no gap separating the particle from the ether in which it is immersed; in the last resort there is absolute continuity. The particles do not act on the ether through a gap; in that sense there is nothing else but contact : the whole cosmos is welded up into a unity, every part connected with every other part. Hence that thesis about contact is both true and false; true if we attend to matter only, false if we attend to the ether also.

Brief statements like that are always liable to this double sense : we cannot really summarize Reality in a phrase : all that we can summarize is one aspect of reality, the one to which we are at the moment calling attention. A phrase is useful for calling attention, for making people think; but we must never use a phrase as a basis of an argument, or set it up in opposition to a fact. Against a fact it is powerless. Phrases may illustrate facts, or may misrepresent them; the only good of a phrase is to focus the human mind.

Now see if the other phrase has any aspect of falsity. " A body cannot act where it is not." Or in other words, every action must take place through a medium. In the material universe this seems to be universally true. But is it true in the vital and mental universe? If we encounter facts which seem to falsify it, we should not deny or ignore those facts on the strength of a phrase; it has no power of confrontation, it forms no basis for an argument, it is a summary of experience, and experience may go beyond any convenient and partial summary : the phrase may have to be extended.

I do not wish to extend it now : I only wish to point out that it may have to be extended or qualified. The question only arises when we seek to generalize the term " body " and apply it also to Mind. Can one mind act on another directly, or otherwise than through a medium? We are now getting into a region of some vagueness. We really do not know how mind acts at all; we know that a thought somehow operates on matter, primarily to all appearance on the matter of the brain, and thus causes physical changes; and we also know that another brain, stimulated into activity by reception of those physical changes through appropriate mechanism, reacts on an associated mind

84

and arouses a corresponding thought; so that indirectly, through a series of physical and physiological processes, one mind can act on another. Even the mind of a dead person can act, if he had left behind a poem or picture or mathematical theorem. It is proverbial that events can be controlled by " a dead hand." Apart from that, the mind of Beethoven or of Newton or Shakespeare is in a real and beneficent sense fully active to-day. But can a mind ever act on another mind directly, without all those intermediate and curious physical methods? The assertion is sometimes made that it can, and the facts responsible for such an assertion are summed up under the head telepathy—mental action at a distance—though what " distance " means in relation to Mind I do not know.

Hitherto we have been denying action at a distance, and saying we must always look for a medium. Are we going to contradict that now and say that mind can act without a medium? That would be rash, and yet it *may* be true, if by further study we find that facts support such a statement. Our denial of action at a distance has no weight as an argument; it is a precautionary measure of policy. We should be wise to seek for a medium if we can, or until we find that such an hypothesis breaks down. We do not know enough about mental action to dogmatize; we must accumulate and study the facts; they must be our guide; a satisfying theory is not yet. It would be utter folly, however, to deny or ignore facts on the ground of preconceptions, because they do not seem to fit into such scheme of the universe as we have already formed; folly to set up against genuine experience our denial of action at a distance.

We do not know the nature of Mind, nor the laws of mental action : they may be quite different from physical laws. We do not know if an active mind need belong to a living person, that is to a person with brain-nerve-muscle mechanism. We have to make experiments and accumulate experience by aid of this mechanism; but we should not let it dominate or hamper us. We may have to enlarge our conceptions to comprehend much more. The region of Art and Beauty and Love and Aspiration is a region higher and beyond anything apprehended in physical science; it behoves us to walk warily and modestly. And it may help us to remember that the greatest men of science have never set up their extensive though still partial knowledge in opposition to the existence of a spiritual universe and direct religious experience.

The Universe, in its wider sense, is infinitely comprehensive; the stellar spaces and the processes of Nature do not exhaust it. These can rightly be regarded as manifestations of some great Reality beyond.

Chapter VI   Electricity and its Action across Space

Chapter *VI* *Electricity and its Action across Space*

What an electric charge is like.

The *whole* mass of any body is just the mass of ether surrounding the body which is carried along by the Faraday tubes associated with the atoms of the body. In fact, all mass is mass of the ether; all momentum, momentum of the ether; and all kinetic energy, kinetic energy of the ether. This view, it should be said, requires the density of the ether to be immensely greater than that of any known substance.—SIR J. J. THOMSON.

When a jar is full of gold or of lead, it does not contain more substance than when we think it empty. —DESCARTES.

THE functions of the Ether we have so far dealt with are :—

(1) those which depend on its continuity, especially cohesion and gravitation, uniting the particles of matter, and

(2) those which depend on its vibration and wave propagation, including all the phenomena of light and radiation.

We have found that this last enables us to arrive at certain properties of the Ether, properties which in an ordinary medium we should call elasticity and density : and we have further said that these have to be accounted for, not in terms of mechanics, but in terms of electricity and magnetism. To proceed further we must know something about electricity, and about what an electric charge is like.

Electrons and Protons

Discoveries of the present century have shown (what had already been dimly suspected by Faraday and Maxwell in the last century) that electric charges are discontinuous, like matter, that they exist as separate particles, although their field or region of influence extends throughout space.

Electric particles or corpuscles are of two kinds, the positive kind and the negative kind (so-called). They came by these names historically, and whether they are appropriate or not remains to be seen : no importance need be attached to the idea of positive and negative, except that they are opposite in sign, and can therefore neutralise each other. Whether they ever do really neutralise each other we do not know; discharge through vacuum is very difficult, perhaps impossible. To ordinary views it is marvellous that things which differ in potential by a million volts can approach within ultra-microscopic distance and retain their charges without loss : yet as a rule they do.

However this may be under exceptional circumstances, the oppositely charged particles certainly attract each other, and when they come very close together they practically blot out each other's field at a distance, so as to form a sort of neutral combination. Particles need not fall to-gether because they attract each other : the Sun attracts the planets, but they do not fall into it, they revolve round it, and their revolution keeps them sufficiently apart; if their revolution was stopped, they would fall in. So it is also with the positive and negative particles : the negative revolve round the positive, and thus constitute a neutral group which we are familiar with as *an atom of matter.* That is what an atom of matter is, and that is what is meant by saying that matter is electrically constituted, or that its properties have to be explained electrically in the last resort.

But we are not yet entering on the subject of matter : we are dealing with the electrical particles, and until we know the constitution of those particles we cannot proceed very far. It is fairly evident that they are not foreign bodies in the Ether : they are probably composed of Ether in a certain special condition. For instance they are subject to gravity, whereas the rest of the Ether is not. Certainly the proton has weight, and probable the electron too, though that is not so certain. They are freely capable of locomotion and are very tractable : the rest of the Ether is not. Electrons have been likened to knots on a piece of string. A knot is not a foreign body on the string, like a bead; it is composed of string in a particular configuration, and yet it is not like the rest of the string; it has a structure and identity of its own. If it is loosely tied, it might be moved about along the string. The analogy is an exceedingly rough one, but it may help. The knots can be localised and can be moved about; but unless you can get hold of one end of the string they cannot be untied : they seem permanent. If they were untied, they would be resolved into ordinary string, and the knot, qua knot, would have gone out of existence, though plainly nothing substantial would have been destroyed. Nothing substantial is ever destroyed; a real thing merely changes its form : it may lose the special properties which depend on a particular form, but the substance remains. Real Substance, like Energy, is indestructible.

So it may be presumed that if an electron could be untied or dissolved, it would lose its properties of electric charge and be resolved into ordinary Ether. No one knows how to do such a thing, no one knows if it is possible; but the fact that electric charges or corpuscles are of two opposite kinds, like plus and minus, suggests that they might, by running together, obliterate each other; not destroying their substance, but losing the peculiarities which they possess by reason of their structure, and ceasing to be electric charges. If such real collision ever happens their energy would be emitted into the ether as a flash of radiation. Some great astronomers think that this may be happening in the giant stars; perhaps even in the sun (see Chapter IV).

Moreover we may presume that if it ever becomes possible to produce these charges, they would always be produced in pairs, never a negative alone or a positive alone, but always both. All electric phenomena may be attributed to the attraction which exists between the two opposite charges and their tendency to rush together. They are always united by what are called lines of force, the lines along which the attraction acts : they travel about with these lines of force attached to them; the lines start from a charge of one sign and end on a charge of the opposite sign. And the electrical phenomena that we observe can be expressed in terms of these lines of force,—as Faraday was the first to show, and as J. J. Thomson has further elaborated. The lines of force are called an electric field, and they represent something going on in the Ether : while their end points, the charges themselves, represent structures or singular points in the Ether of at present unknown character.

To proceed further we must make hypotheses. Hypotheses are things to be held lightly and tested : they constitute working clues, they may have more or less truth in them, but until they are confirmed they do not constitute a theory : hypotheses are useful as far as they go.

The forces and the laws according to which they act are not a hypothesis but a fact : and electrical phenomena can be worked out, as Newton worked out the phenomena of Gravitation, without understanding what the thing dealt with really is. Yet the human mind inevitably seeks what everything really is, and during the search makes hypotheses.

We can now summarise briefly what we know. The two oppositely charged particles, the negative and the positive, are called respectively an electron and a proton. They are both exceedingly minute : and there is a sense in which their size has been measured. They are far smaller than atoms, incomparably smaller, the smallest things known : even if there were a hundred or a thousand of them in the atom, they would not be in the least crowded, there would be plenty of empty space. Different atoms are now known to be composed of a different number of electrons, and by their different number and grouping they constitute the different chemical elements. The atoms of all the chemical elements are built of electrons and protons and of nothing else.

I do not want to deal with matter now, only with the Ether; but it is necessary to realise that if the ordinary masses of matter are composed of electrons, and if the electrons are so small that most of the matter is mere emptiness, while yet we know that matter may be as dense as lead or gold or platinum, it is evidence that the particles themselves must be of enormous density, and that that density must belong to the Ether of which they are made. We could not have told what the density of the Ether was but for these structures in it; but by aid of these structures and their behaviour, an estimate has been made of Ether density; and on the view which I advocate it is enormously denser than any form of matter, just about a million-million times that of water.

It used to be thought that the Ether of Space was exceedingly rare, much rarer than air, rarer even than the residual air in a vacuum tube. But if matter is made of Ether, that is to say if matter is built up of electric particles, and those particles are composed of Ether, the view that Ether is rare is untenable. It is the densest thing known : there can be nothing more massive than Ether : for, being a continuum, it is incompressible. No part of it can be denser than the rest. The knots on a bit of string are no denser than the rest of the string; and by weighing a knot and determining its size, we could estimate the density of the string. In the parable or rough analogy, we must suppose that the unknotted parts of the string are inaccessible to us and escape observation, we are only aware of the knots, we have to infer the rest of the string. We infer that it is just as massive as any modified portion of it can be. The modification into matter does not increase the fundamental density. This, if true, will be as Descartes surmised (see quotation at beginning of chapter). In other words, the apprehensible modification of Ether that we call matter need not be any denser or contain more real substance than the rest.

If Space is completely full of substance, and if that substance is of great density, a difficulty has sometimes been felt as to the possibility of locomotion, whether indeed motion was possible in a plenum. The difficulty is not a real one. Resistance to motion is due to viscosity, not to density; and the ether certainly has no viscosity : it is not at all like treacle, it is perfectly limpid. Density is no cause of friction, but it is a cause of inertia, and inertia, is just what moving bodies exhibit. While as to motion through a substance of which Space is already full, one need only point out that a fish can move freely in the depths of ocean. This difficulty is imaginary ; or, in so far as it is genuine, it can be countered in other ways.

Suffice it to say then that arguments too elaborate to be reproduced here, but which I have published elsewhere, give the density of Ether compared with water, as 1 followed by twelve 0's,—of that order of magnitude.

Let us grant this as a working hypothesis, without further argument : on that basis we can proceed.

The rate of propagation of ether-waves depends on the ratio of elasticity to density. It is that which determines the velocity of light; but the velocity of light is known : so if the density is granted, the elasticity is known too.   It must be expressed as 1 followed by thirty-three 0's.   It is excessively great !

But what kind of elasticity is it ?   When we were treating of waves we saw that there were several kinds of elasticity : a gas had one kind, a solid had another.   What kind is the elasticity of Ether ?   We do not know; probably it is a new kind.   Attempts have been made to explain it by the vortex motion of a perfect fluid : Lord Kelvin made such attempts : the calculations were very difficult : the result is not wholly satisfactory. We have to admit that we do not positively know the kind of elasticity it is.   The simplest plan is to assume the simplest case, that it corresponds in some way to the incompressibility : in other words that it represents or involves or has the same value as the pressure.   I shall make this further hypothesis, that everywhere the Ether is under a pressure represented by 10^{33}.

Granting that pressure, we can go forward again; and the existence of an electron can be understood : we can reckon that it is this pressure which holds the charge together against the mutual repulsion of its parts. It can only do so if the electron is of a certain size : the size can be calculated, and it comes out just the same as the size determined by experiment. The charge of the electron is in equilibrium, on a sphere of known size, under this enormous etheric pressure.

On this view the existence of an electron can be fairly understood. Can the existence of a proton be understood too ? No : there we are in a difficulty. The proton is more massive than can easily be accounted for : and why it is more massive we can only guess : indeed at present we can hardly guess, or at least the guesses are not very satisfactory. That remains at present an outstanding puzzle : the question is one that has hardly yet been faced. One guess is that the electron is hollow, like a bubble, that it has an electric field which by itself would cause the bubble to expand, but that it is kept in equilibrium and of a certain size by the etheric pressure. On this view there is no substance in its interior; in itself such an electron is not massive at all, its apparent mass is due to its

101

electric field and to nothing else. Whereas the interior of a proton, instead of being hollow, may be full-filled with extra ether; all that which was removed from the electron being crammed into the proton, so as to account for its great massiveness or what we may call its weight. A proton is more than a thousand times as heavy as an electron, about 1840 times by direct measurements; and what is called " the atomic weight," or the weight of an atom, depends almost entirely on the weight of the protons it contains. The hydrogen atom contains only one, the helium atom contains four, the lithium atom seven, the oxygen atom sixteen, and so on—in accordance with the list of atomic weights long empirically known in Chemistry, the heaviest being uranium, which contains 238. The atomic weights are certain enough; the number of protons in a specified atom is fairly certain also. What is not known is why the proton has such a weight, and why the weight of an electron is so much less. In every other respect the two charges seem equal and opposite : electrically they *are* equal and opposite : materially (whatever that may mean) they are not.

A charged body has additional mass. The whole mass of an electron is accounted for by its electric field. Not so with a proton. That has mass over and above its electric charge,—mass presumably belonging to its ether content. It seems extra full, while an electron seems empty. Electrically they are equal and opposite. Materially they are as 1840 to 1. A queer number this 1840. It is only known roughly, it is not known to be an exact number; but it probably is, and if so it must have a deep meaning—a meaning not yet begun to be deciphered.

We are safe in saying that the weight of matter depends on the protons, that is the positive units, which go to form the nucleus of the atom, while the chemical properties of the atom depend on the electrons which circulate round the nucleus. These planetary electrons are active and energetic and produce conspicuous results : they characterise the atom by its spectrum; they confer on it its chemical properties; but they add to its weight hardly at all. It is a curious state of things, but the evidence for it, so far, is good.

For the rest we can only say that an electron *behaves* as if it were hollow, with no mass other than its electric field; while a proton *behaves* as if it were crammed with matter. We might guess that its nucleus contained the substance removed from the nucleus of an electron; thus constituting an individualised portion of the main ether, endowed with locomotive powers, and inseparable from its electric field. This is an idea that must be elaborated elsewhere; we are now near or perhaps beyond the confines of knowledge.

Reverting to our parable at the end of Chapter III about explorers plunging through an obscure wood, we can realise that the work of our Elders was all preparation: our scientific ancestors tried every other road, and failed, and by their failure pointed us out a new way. They dimly perceived two obscure regions ahead, which we have begun to enter—the region labelled " relativity " and the region labelled " quanta "; their labour was by no means lost, it is all part of the great scheme. They cleared the ground to an amazing extent, their achievements will all fit in to the work of present and future generations, but it may be necessary to lose sight of some of their work for a time. They have erected gorgeous dynamic structures, massive buildings, firm and solid. We, immersed in the thicket, cannot hope to build as yet : the time for building will come, the time for exploring is on us : we are bound to follow the gleam.

o

Active forces are conserved in the world. It has been objected that two soft or inelastic bodies, when they collide, lose part of their force [energy]. I answer that this is not so . . . the loss ensues only in appearance. The forces are not destroyed, but dissipated amongst the minute parts. That is not as if they were lost, it is like the changing of large coins into small ones.—Leibnitz.

One of the most important things that we know about electrons and protons is that they can never be either brought into existence or put out of existence : they can only be moved about from place to place,—they are in that respect like any other form of matter. Parenthetically, we find ourselves rather singularly in accordance with the electrical views of Benjamin Franklin long ago :—the electron is all electricity, the proton feels more material. He only admitted one kind of electricity : his other kind was " matter." Charge is transfer of one kind. Electrons are easily transferred from one body to another; they can be accumulated so that one body has an excess, but some other body must have an equal defect. An extra number of electrons gives a negative charge to a body : anything with an excess of unbalanced protons is positively charged.

But it may be said, surely when a thing is discharged (that is, when one body sparks into another, as when you connect the two coats of a Leyden jar), surely the two charges disappear ! Yes, they " disappear " : that is the right word ; they are no longer apparent ; but they have not gone out of existence. They have re-grouped themselves ; the two opposite charges have got close together, they have as it were re-united, they are mutually satisfied, they make no more disturbance or strain, their lines of force have as it were shut up or become exceedingly short, they have neutralised each other so far as outside effects are concerned ; but they have not obliterated each other, their existence is not terminated : they are there ready to be separated again, though the oppositely charged bodies have become neutral.

We never really generate or destroy electricity : what one body has gained, another has lost, and when the balance is restored there is equilibrium. We produce or display electric charge simply by transfer, simply by making manifest what was previously existent. We pull out infinitesimal lines of force and make them extend across perceptible space : we thus in a way generate an electric field; but the field was pre-existent, though infinitesimal and non-apparent before we displayed it. If we liken the lines of force to elastic threads, they are elastic threads of infinitesimal length, capable of being stretched ad libitum, without limit; the lines never snap, nor do they ever shrink up into absolute nothingness. To make this statement quite safe, we ought to say that they are not " known " to do such a thing in any of our experiments, however violently we treat them. Whether they ever shrink to nothingness of their own accord, remains to be seen : such a discovery has not yet been made.

What we have now to learn is that something of the same sort is true of magnetism : we never really generate magnetism any more than we generate electricity. We can make a magnet, just as we can charge a body; but the magnetism was there beforehand, just as the electricity was : we have only made manifest what was pre-existent.

So far there is similarity : but now there are differences. We cannot charge a body with one sign of magnetism, say the north-seeking kind, leaving another body charged with the south-seeking kind : every magnet has an equal quantity of both kinds, though instead of being close together they can be separated by the whole length of a piece of steel.

That is an old-fashioned method of regarding magnetization, and has been known for centuries, but the inner meaning of it has not been so thoroughly known.   There is a fundamental difference between electric and magnetic lines of force, a difference quite definite, though perhaps not easy to specify completely.   The main difference is this :   an electric line of force has two ends, a positive end on a proton, a negative end on an electron. A magnetic line of force has no ends, it is always a closed loop : we may not think of a magnetic line of force as analogous to an elastic thread reaching from one particle to another and pulling them together; it is more like an indiarubber ring, it is looped or closed on itself, it has no ends.   But the loop is one that can be stretched, it can be expanded so as to enclose a big area, or it can shrink up

110

until it is too small to be appreciable. It never shuts up to nothing; it never goes out of existence. If you call these loops magnetism, we never generate magnetism; we open it out. The act of magnetization opens a loop out, and, being elastic, it is always trying to close up. When a loop encloses two bodies, and then shrinks, it pulls them together as if they attracted each other. That is magnetic attraction. Rather like an india-rubber ring holding things together.

An inert piece of iron is full of such infinitesimal loops : the act of magnetization opens them out and makes them perceptible. Part of their course then lies in the air : but the air has nothing to do with them, they do not really exist in the air : they exist solely in the Ether. And the iron is not necessary either : it may have an extra number of the loops, and without iron or some similar substance we might possibly not have discovered the magnetic loops, or not discovered them so soon; but they are there all the time, everywhere.

The Ether is full of them : and it so happens that in or near some large bodies, like the earth, some cause has already opened out a few of those loops; so that the earth behaves like a permanent magnet. What has magnetized the earth we do not completely know; but it was owing to that fact that the Ancients were enabled to discover magnetism. And some substances, even when dug out of the earth, have the power of retaining these loops in an opened-out condition : in other words, some ores of iron are natural magnets. They are not very strong, —artificial magnets can be made much stronger,—and, as the cause of natural magnetism is not known, it is not the most instructive kind for serious contemplation.

It was long ago known however that the power of retaining the loops in an opened-out condition was possessed by steel, and many other substances in a greater or less degree; and there were rather blindfold methods also known by which one magnet could excite another. The natural idea would be that the magnetism of one was transferred to the other, and that what one body gained, the other lost. That is not so. It is so in electric charge; it is not so in magnetism. The magnetic influence, or so-called " induction," is exercised without loss : one magnet can magnetize any number of others without itself being any the poorer; there is no limit to the amount of magnetization that can be produced, once it is started.

Analogy with Life

In this respect it is rather like Life. An oak can produce a large number of acorns, and each acorn has the potentiality of producing an oak; and so on without limit. Wherever life exists, it can increase prodigiously; not at the expense of the original source, but as if there were an infinite reservoir from which it could be drawn. The analogy is striking, even if not helpful; for though we only know a little about magnetism, we know much less about life. We do not explain the unknown by the still more unknown, but we can draw attention to analogies and similarities, in spite of immense and important differences.

There is this further similarity, however, that production of more magnetism, or more accurately the display of pre-existent magnetism, is not accomplished without expenditure of something. It is not magnetism which is expended, but energy : energy must be expended, that is, work must be done, in order to produce more magnets; and if energy is not expended, nothing happens. So it is with life. A plant can produce innumerable seeds, and thus give rise to innumerable plants, but not in the absence of energy, not without the rays of the sun : there must be available energy, and that is consumed in the process. We live in such a stream of energy, coming from the sun that we almost forget it : yet it is essential to every movement, it is essential to the display of every form of life. Life may exist in some imperceptible form unknown to us, but without radiant or etheric energy, it could not enter into relation with matter, it could not grow and develop and become conspicuous. Energy may not be necessary to abstract Life, but it is essential to the display or manifestation of life in matter.

What Life is we do not know : what magnetism is we are only beginning to learn. But the two have this in common : they may be dormant, imperceptible, inactive; by means of energy they can be displayed, and displayed in unlimited amount. There is, as it were, an infinite amount available; or if not infinite—that is without limit—the limits are beyond anything that we have detected or imagined. We cannot say much about Life (at least I cannot), but I can say something more about magnetism, which is clearly a much simpler thing, and I will proceed to make some assertions about it.

We not only have the power of passing it on, by help of a piece of magnetized iron ore which we pick up; we have the power of magnetizing a body *de novo*. We cannot generate the loops, it is true, but we can open them out, even if none already opened out were given us. We do not require a permanent magnet in order to magnetize bodies. What do we require? We require a moving electric charge; that is we require to get an electron and a proton separated from each other, and rush one of them along. We generate magnetism by means of electricity : the two are remarkably inter-related : they are by no means the same thing. Electricity can generate magnetism, with the aid of energy expended : this was discovered by Arago early in the Nineteenth century : and a few decades later Faraday made the tremendous discovery that magnetism could generate electricity,—also by the expenditure of energy. And now every dynamo is doing it.

But these terms are not quite accurate. When I say that electricity can generate

magnetism, I am speaking popularly.   More
accurately, an electric current can open out
pre-existent magnetic loops and thus generate
a perceptible magnetic field.   And when I say
that magnetism can generate electricity, I
ought to say that a magnet enables energy to
generate an electric current, that is to set
electrons in motion.   Motion, mechanical
motion, is the link between the two things,
electricity and magnetism : and it is by aid
of mechanical motion (which is not an etheric
process, pure and simple, but a material one)
that electricity and magnetism in combination are able to interact with matter; and it
is thus that they have come within our ken.
We realise electrical and magnetic phenomena
by the motions which they are able to produce; we have sense organs for the purpose.
Our physiological organism (in some mysterious way, to which we have grown so accustomed that we fail to realise the mystery)
enables us to perceive mechanical motion.
We can see pieces of matter moving.   When
we see the gold-leaves of an electroscope
diverge, we know that they are electrically
charged : when we see a galvanometer deflect,
we know that there is an electric current in
its neighbourhood : when we see two pieces

of iron or steel jump together we know that they are magnetized.

We have a further most extraordinary sense. When we look at the flame of a fire or the filament of a lamp, or at the moon or a star or a landscape or indeed any visible object, we know—at least some of us have learnt to know—that there is an electro-magnetic disturbance in the Ether, travelling at a known pace, reaching our eye, and in some curious way stimulating the optic nerve. The sensations we experience directly : the meaning of those sensations, the causes which have excited them, the processes to which they are really due, are matters for inference, that is for scientific discovery. The senses give us the result, they do not tell us about the mechanism : it has been no easy matter to discover the mechanism, and there is much still to be discovered about it : our explanations are bound to be dependent on the present state of scientific knowledge; we may be sure that posterity will know much more. Meanwhile we can try to learn what is already known : and in so far as our knowledge is inaccurate as well as incomplete, we must be always ready to correct it. We are certain that our knowledge will have to be enlarged : our hope is that it is accurate as far as it goes. But even of that we must not be too certain; and in so far as some of the problems are difficult, we must be prepared for differences of opinion among skilled and expert investigators.

Discussion and friendly controversy are useful in Science. What is not useful is the kind of political controversy and international greed which distract attention from the beneficent processes of nature which humanity can hope to understand and utilise. Persons so afflicted are sometimes seized with madness and seek to apply the knowledge already acquired to the arts of mutual destruction. If the different sections of humanity could only settle down to peaceful co-operation, ascertaining and interchanging knowledge freely, without wasting energy in barbarous conflict, there is no end to the possibilities in front of us. We have learnt so much, we are always learning more; and the utilisation of the energies of the Universe is one of the functions of civilised man. We live among difficulties, our life is bounded by a struggle; but it may be a co-operative and not a competitive struggle.

Knowledge in this respect is like life and

magnetism : there is an unlimited reservoir from which to draw, and the imparting of knowledge does not lessen the amount possessed by the imparter, it is transferred without loss, though doubtless with the expenditure of some energy. Knowledge grows from more to more. By diffusion it is increased : what one gains another does not lose. Some energy and skill are needed for broadcasting ; but it is an operation that can be conducted without loss. If the transmitter suffers at all, it is only his bodily energy which is consumed. A magnet which has excited other magnets may be even stronger than before : life which has excited other life may still be vigorous. There may be temporary fatigue : a shrub may be temporarily exhausted by producing seeds, but it soon recovers : fatigue is natural to material organisms and to any expenditure of energy ; but recuperation follows, and next year it is ready for a fresh crop.

Certainly in the case of a physiological organism, fatigue may be pressed to excess : exhaustion, old age, and death lie in wait for bodily organisms : it is not they which are permanent. A magnet may be knocked into inefficiency by hammering ; its loops shrink up into apparent nothingness, but they have not gone out of existence. A loose tile from a house-top may destroy the mechanism of thought, and damage irretrievably the working of an organism : thereby life and mind will disappear from our ken. But if the analogy holds, they have not gone out of existence. So far as we learn from Science, nothing goes out of existence; it only changes its form, it may become inappreciable to our senses, it may to all appearance cease to be. That is the Appearance. What is the Reality ? That question must be answered by Science. In the case of magnetism and electricity it has been answered; in the case of life and mind it is in process of being answered,—but of course not without controversy. Friendly discussion and controversy is to be welcomed. In this, as in all other matters, we must learn from the facts.

Chapter VIII Electromagnetism

Q

Chapter *VIII*   *Electromagnetism*
How radiation is generated : and its problems.

There is weighty argument to be adduced in favour of the ether hypothesis. To deny the ether is ultimately to assume that empty space has no physical qualities whatever. The fundamental facts of mechanics do not harmonize with this view. . . . According to the general theory of relativity space is endowed with physical qualities; in this sense, therefore, there exists an ether. According to the general theory of relativity space without ether is unthinkable; for in such space there not only would be no propagation of light, but also no possibility of existence for standards of space and time. . . . But this ether may not be thought of as endowed with the quality characteristic of ponderable media. . . . The idea of motion [locomotion] may not be applied to it.—*Sidelights on Relativity*, by PROF. EINSTEIN.

Every disturbance of the aether, including radiation as one type of disturbance, is originated by translatory motion of electrons through the aether. . . . The æther is a perfect fluid endowed with rotational elasticity.—SIR JOSEPH LARMOR.

A high speed electron has additional mass by reason of its motion. When it is stopped, this mass is thrown off as radiation. This will be a quantum, greater or less according to the speed : its amount can be reckoned; it should correspond with the energy of the voltage-drop that produced the motion. When radiation throws off an electron, this mass is restored to it : and that determines its speed.—O. J. L.

It is well known that magnetism can be produced by electricity, or more accurately that a magnetic field can be excited by an electric current. We have now to try to enter on the more intimate detail, of the process.

When we were speaking of electrons as the basis of electric charge, we were not satisfied to accept the electrons as units given without explanation : we tried to speculate as to their nature, and see if we could imagine some structure in the Ether which would account for them. Let us try to do something of the same sort with magnetism : we may not succeed, but at least we can state certain undoubted facts. What I say about an electron must be true of a proton also, with change of sign; but it will be sufficient if we attend to one : and at present we know less about a proton than about an electron.

An electron at rest has nothing magnetic about it : it has a field of electric force; lines of force radiate from it in all directions : it seems to have a centre or nucleus in a certain locality, but its lines of force stretch out everywhere. An electron is capable of locomotion ;—locomotion is one characteristic

124

of every form of matter;—it is doubtful if the unmodified ether is capable of locomotion, at any rate we don't know how to set it moving. But we do know how to move an electron, its field of force gives us a handle : it is easy to move, it is the most mobile thing we know : it has extremely little inertia, and the smallest force will set it going very quickly.

When it moves, what happens? Its line of motion is surrounded by magnetic rings : it, as it were, threads magnetic loops. It does not generate those loops, it rearranges pre-existent loops : its line of motion is surrounded by magnetism, as an umbrella might be surrounded by an indiarubber ring. The faster it moves, the more those rings open out, the bigger they grow, until they become conspicuous. A moving electron is surrounded by a magnetic field.

This has important consequences, extra-ordinarily important : and if we knew all that was happening, and exactly what the rings were like, we should know a great deal more than we do. But we know some of the consequences : one is that an electron displays the fundamental property of matter, the property of inertia, the property of going on until it is stopped : once set in motion, its motion will continue; it will continue to move in a straight line with uniform velocity, unless deflected by external agency. An electron has a certain massiveness; although so minute it is not a geometrical point, it is a substantial thing : it can deal a blow to an obstacle put in its way. It has a certain amount of energy; every moving body has energy : the energy of a cannon-ball or of a motor-car is conspicuous, you cannot stop it dead without violence : put an obstacle in the way, something happens.

The energy of a moving electron can be calculated. If it is stopped, what becomes of that energy ? Well, what becomes of the energy of a cannon-ball when that is stopped ? That is more complex : a great moving mass is not a simple thing like an electron. We know that the target is damaged, that the air is disturbed in such a way as to cause waves of sound to travel out from the impact, and that if the blow is violent enough, the ether is disturbed too, and there is a flash of light. Moreover all the molecules are set quivering : in other words a lot of heat is produced.

Does anything of the same sort happen when a quickly moving electron is stopped by a target, that is by some heavy atom which gets in the way? The energy must go somewhere. When the motion is stopped, the magnetic rings cease to be. What becomes of them? Do they go out of existence? Certainly not. Do they shrink up to infinitesimal size, as they might if the electron were checked gradually? No, if the impact is violent enough, the rings do not shrink, they rather expand : the electric and the magnetic fields, which had previously existed quietly together, now combine into an Ether disturbance. They travel out as waves, not a series of waves but a pulse, a shell of wave, rapidly expanding with the speed of light. And they carry with them the momentum of their origin. The moving electron had extra mass,—temporary matter, —and that extra mass, when it stops, is converted into a quantum of radiation.

Such wave-like shells were discovered by Röntgen; they are known as X-rays. It is a most instructive phenomenon, one that can be followed in some detail. The quicker the electron was moving, the stronger and sharper will be the shell of radiation. If an electron had been travelling at what, for it, must be called a moderate speed, a few thousand miles a second, it emits what are called "soft X-rays," *i.e.* the kind which are not very penetrating, which can be stopped or absorbed by fairly soft tissues : whereas if it had been moving say a hundred thousand miles a second, or something approaching the speed of light, the X-rays will be exceedingly penetrating, able to go through bone and iron, what are called "hard X-rays." Thus we are getting on familiar ground. The energy of the wave will be comparable with the energy of the electron before its stoppage; a quantum of energy passes into the Ether : there is no waste in sound, and not much in heat, it nearly all goes out as radiation.

Is the electron destroyed in the process? No. The electric charge remains, and has to be carried away; wires are provided for the purpose : the target must not be insulated, or it will gradually get charged by the torrent of electrons and decline to receive any more. The charge itself is uninjured : whence then comes the energy? It comes from the magnetic field surrounding the moving electron, and surrounding it only

while it is moving : this it is which is broken
up and destroyed, or rather not destroyed,
but distributed as a wave into space, turned
into an etherial quiver. The process is
typical of the production of radiation generally : there is probably no other radiation
except that which is produced by the impact
of electrons and protons. Electric charges
clashing together excite radiation : this is
probably true even in the comparatively
gentle chemical processes of combustion and
flame.

But now comes an important though
perhaps rather difficult point. The magnetic
field surrounding the electron possessed
energy; it possessed momentum, it enabled
the electron to continue in motion, it is
responsible for its inertia; it probably contains the clue to the meaning of inertia.
In what form is the energy in a magnetic
loop or ring? We do not know : we can
make a surmise, based on certain evidence;
but experiment has not yet answered. The
great mathematician, Sir Joseph Larmor,
who holds the Chair of Newton at Cambridge,
has surmised that round a magnetic loop
the Ether is circulating,—not necessarily
quickly, but circulating,—as a curtain-ring

might be spinning in its own plane, or like the rim of a spinning-top or fly-wheel; and that the energy is represented by this circulation. I said that the Ether was not capable of locomotion, or at least that we had no means of getting hold of it and moving it from place to place. That remains true, but a spin is not locomotion. There is no locomotion about a spinning-top, if it is merely spinning; it has gone to sleep, it is stationary. Such motion is sometimes called "stationary motion"; it is not apparent unless you try to stop it. When you try to stop it, something happens : bring something in contact with the edge, the top flies away. Stop the magnetic circulation in a loop, it too flies away,— not like the top, but by expansion; thereby a disturbance is generated in the Ether which we call either X-rays or light, according to circumstances.

What generated this spinning motion in the Ether? What was the origin of these spinning loops? No one knows : Science cannot answer the question of " origin " or genesis. We have no means of generating such loops : they are there : all we can do is to open them out and make them apparent.

They can be opened out quietly, as in an electro-magnet; or they can open out violently, as in radiation.

It must be understood that the loops are a reality. What is going on in them is a speculation : it still remains for experiment to ascertain; to confirm or contradict our guess. One way that I have suggested of verifying the supposed circulation is to make a very strong magnetic field, and send light along it, first one way and then the other, or rather both ways simultaneously, and see whether one is accelerated and the other retarded. The experiment is a difficult one, and has not yet been adequately made. I have indeed made the experiment with considerable care but with insufficient power, and so far without result, that is without detecting the circulation. But there are good reasons for that; if the Ether has the density which I attribute to it, the circulation in any artificial field is almost too slow to be detected. The experiment requires to be repeated on a very much larger scale, if it is to give any result. Meantime we must leave this part of the subject as a speculation, and say no more about it here.

We are sure of this,—that radiation is

produced by atomic or electronic collision; *i.e.* by sudden contacts at high speed. And we have already studied the nature of 'contact' (see Chap. IV) and found that a collision is of an astronomical or etheric order, that atoms of matter are never in contact, there is always a sort of elastic cushion which keeps them apart, and that by interaction of etheric fields radiation is generated, and generated at the expense of a temporary kind of matter produced electromagnetically by motion.

Surmises

I am here only touching the fringe of a vast subject,—the connexion between ether and matter, and the temporary kind of matter which is consumed in producing radiation, and on which the size of the quantum depends. We are on the borderland of knowledge, and must walk warily.

There are strange unexplained facts which we see looming ahead of us. Radiation has been studied for a century in the full light of a wave-theory, but it still contains outstanding puzzles. It is emitted and absorbed in packets, in quanta; radiation simulates some of the properties of matter, it carries momentum and exerts pressure; who knows what its exact relation to matter is? Sometimes we feel as if radiation were a half-way stage between ether and matter — something half-way between the free and the modified ether. Matter is discontinuous. Is radiation discontinuous too? Is light discontinuous? Is it a kind of matter which is bound to travel at a fixed speed, and is only tolerated by the ether at that speed; or only otherwise tolerated if it be modified, as when its energy is imparted or converted into an electric charge? Strange notions these, but we have to get used to strange notions. Whatever the truth may be it will be profoundly significant when we grasp it. The singular difficulty and the intense interest of the problems before us are only equalled by the ingenuity with which they are being attacked.

Meanwhile the only way to progress safely is to study the facts, especially the obscure facts. Somehow light bears traces of its origin from quivering electric units; and it exhibits those traces forcibly when it encounters similar units. Matter excites radiation at its own expense : does radiation ever, anywhere, give rise to matter? The eternity of the cosmos seemed at one time doubtful, because of the dissipation of energy: now there is some glimpse of a way out. Matter tends to fall together gravitationally: but radiation tends to spread to the confines of the universe. And yet, however diluted, it retains its vigour. When quanta are thus isolated, what are they? What becomes of them? There is some deep meaning even in the speed with which light travels: no greater speed seems possible. In that speed, and in the newly presented puzzles about the connexion between ether and matter, and about the nature of radiation,—problems which are beginning to take definite shape,—we are down among the foundations of material being. We stand as it were enthralled by the revelation which is dawning upon us. Confronted with a majestic vision of Reality, we—like those other explorers on their first view of the Pacific Ocean—have

" Look'd at each other with a wild surmise——
    Silent, upon a peak in Darien."

Chapter IX   Matter as One of the Forms of Ether Energy

Chapter IX   Matter as One of the Forms of Ether Energy

What an atom is like, and what becomes of it.

It is incorrect to try and explain matter as something real, and force as a mere notion to which nothing real corresponds; both are abstractions from the real, formed in exactly the same way. We can perceive matter only through its forces, never in itself.— HELMHOLTZ, *Erhaltung der Kraft.*

So far we have dealt, after a fashion, with the Ether and its functions; instancing among those functions gravitation, cohesion, light and radiation, electricity and magnetism. Are there no other functions that have to be considered? Yes, there is one very important one, the constitution of matter itself.

Electrons and protons are the building stones of which matter is made. The atom of matter is composed of them, and all matter is composed of atoms. Electrons are evidently composed of Ether, because whatever mass they have is represented by the energy of their electric field, which is certainly an etherial phenomenon : and apart from this field they seem to have no other existence; they are electric charges and nothing else. We cannot make a similar statement about a proton, because we do not know enough : for that we must wait. The things that I might attempt to say about it are too speculative at present, and would probably be wrong; this at any rate is not the place to say them.

Meanwhile we know that an electron has mass represented by its energy. We also know that a moving electron is more massive than one at rest, and that as its speed increases, its mass and energy slowly increase too, preserving constant proportion to each other. Matter is turning out to be one of the forms which etherial energy can take— a very curious and permanent form, not easily changed into other forms; at least not the whole of it : some of it is. The part of it which is easily changed is the extra mass acquired by a moving electron : this behaves like additional matter, but not like permanent matter. When an electron is stopped, this temporary matter disappears : it does not vanish into nothingness, it is changed into radiation; it is as it were shaken off or shocked off from the electron, and travels out into space as a *quantum* or individual splash of radiation with the speed of light. It may be actual visible light, but it must be radiation of some kind, whether it affects the eye or not : in its best known and simplest form it is an X-ray.

All radiation is produced by a sudden change in the motion of an electron; and the wave-length depends on how fast the electron was moving, and how quickly it is stopped. The electrons circulating round an atom have the power, the peculiar power, of dropping from one orbit to another, every now and then. Their orbits are not fixed like the orbits of the planets : it is as if Mars could suddenly drop to the orbit of the Earth and begin circulating there. That is one of the peculiarities of atomic astronomy : one in which it differs from celestial astronomy. In celestial astronomy the only objects liable to drop from one orbit to another are the comets : these may be captured by one of the planets and be thus dropped into a totally different orbit, being at the same time apparently pulled to pieces and converted into a swarm of meteors. We will not pursue that further, because there is nothing certainly known in the atoms to correspond with that.

Atomic astronomy is in some respects simpler than celestial astronomy; for all the planets or satellites revolving round the nucleus are similar to each other, identical in everything except speed and position : and their orbits, though very complicated, are by no means distributed at random : only some orbits are possible. There is a curious kind of discontinuity in the immediate neighbourhood of a material nucleus : the satellite or planetary electrons can occupy certain positions and no others : they can drop from one of these possible positions to another, and when they do so they emit energy in the form of radiation. It is as if they were stopped at the new orbit and thrown into shivers by the stoppage, so as to emit X-rays, or it may be light. The kind of radiation they emit depends on how far they have dropped and where they drop to. If they drop from a long way off (say a millimetre) into a position close to the nucleus, they get up a big speed, shiver quickly, and emit ultra-violet radiations or X-rays : if they drop only a little way, and remain at some distance (say a ten millionth of a millimetre) from the nucleus, they emit what is called infra-red radiation, something below the visible range. Some of the radiation can be seen; nearly all of it can be photographed; and that is how we learn about it. Each drop nearer to the nucleus corresponds to a line in the spectrum; and by analysing the spectrum the structure of the atom has been made out. We have more to learn about this process of drop, and why certain orbits only are possible; but all observers agree that radiation is due to a sudden drop, though they may differ as to what the drop means and what kind of thing it is.

The process is a reversible one : not only is radiation emitted, it can also be absorbed; and when radiation is absorbed, the electron is jerked up again. How far it is jerked up depends on the kind of radiation : a hard or high-frequency X-ray would jerk it up a long way, even out of the atom altogether : a soft X-ray or visible light would jerk it up from an inner orbit to an outer one. Whence in due time it might drop back again, giving out exactly the same energy as it received when it made its jump.

All these details can be followed, and are

studied by the great spectroscopic analysts who are now at work. The spectrum has thrown a flood of light on inter-atomic processes. And by " the spectrum " I mean the line spectrum, the discontinuous spectrum, the kind of spectrum which is full of law and order; not a continuous spectrum, which is a mixture of a number of different kinds, but the spectrum due to a simple and definite process. It must suffice here to call attention to this great subject, and leave it for the future elucidation of the workers, not because too little, but because too much is known about it. And yet, in spite of this knowledge, our understanding of it is incomplete. Everything relating to the " quantum," or individual splash of radiation, is rather puzzling : we have only partially got the clue.

What concerns us now is the fact that matter and energy are equivalent to each other; I do not say identical, they are different forms of energy. The important thing is that they *are* forms of energy, so that when one disappears, the other makes its appearance. Energy is always protean in form : at one moment it can be mechanical strain, as in a stretched bow; at another moment it

142

is visible motion, as in an arrow; at one moment it is in the form of compressed air or steam, at the next it is the revolution of a fly-wheel; at one moment it is a raised 20-ton weight, at another moment it is the heat and noise when the weight has fallen and crashed; at one moment it is the energy of an electric current, that is of magnetism, at the next it is the motion of a street car, propelled by that magnetism; at one moment it is an electric charge, at another it is the heat and light of a spark. All this has long been known, but we are now surprisingly able to say that at one moment a portion of energy is matter, mass, inertia, momentum; and the next, it is away travelling in the Ether as radiation. Moreover, energy which is now radiation can make an electron jump, and thus constitute an electric current, with all its intangible and protean properties.

Are we able to say that radiation can actually generate matter,—not only the temporary form of matter associated with electromagnetism, but the permanent form of electrons and protons? Here we must hesitate and proceed warily: we do not know that at present. There are some who think it probable; but all are willing to hesitate and seek the aid of further experiment. The birth of permanent matter would be a great discovery. Attempts have failed, so far.

But we can proceed, with some hope of getting an answer, to ask the converse question, Do an electron and proton ever clash together and obliterate each other in a pulse of radiation? Can matter, as we know it, radiate itself away and cease to be matter, until at some future time perhaps it is reconstituted? We do not yet know for certain—we only suspect it : what we know is that the permanent units of matter, when in motion, have more mass than they had at rest; and that this additional mass *is* convertible into radiation. This is happening continually, and accounts for all the light and radiant heat with which we are familiar.

In our laboratories, electrons can drop towards the nucleus. Do they ever drop into it? Or must they stop and take up some orbit short of the nucleus? We cannot answer that question yet. There are some who think that though the process does not as yet go on in our laboratories, under the conditions of temperature and pressure there available, yet that in the extravagant conditions of pressure and temperature which exist inside the stars, especially the giant stars, the process may be occurring. The radiation of such stars is tremendous, and it goes on for millions of ages, without apparent diminution. The Sun is rather a small star, but it is known to have been radiating for millions of years, perhaps for thousands of millions : there is no limit to time, time past or time future : the Universe is a going concern, and time seems to be infinite. How can we account for all that radiation ? Whence comes the energy which a body like the Sun is constantly emitting, and which some stars are emitting thousands of times faster ? And what has become of it all ? Here are two questions. We are not yet ready to answer the second; but Astronomers believe they can answer the first.

They say that not only the atoms must be falling together, they suspect that the very constituents of the atoms must be falling together. I do not say that it is true; no one positively asserts as yet that it is true; but if it were true, it would account for the energy. Not only the temporary and electromagnetic matter is disappearing, some of the permanent matter may be dis-

T   145

appearing too. Matter may not be so permanent as we think. It seems permanent
enough under terrestrial conditions, but we
can see how the destruction or transformation
of matter could produce radiation; we are
beginning to think that that is how most of
the stellar and solar radiation is produced.

We can already say, pretty definitely,
that wherever radiation is produced, matter
disappears. Whether it be only the temporary form, or also what has hitherto been
thought of as the permanent form, is the only
question. The radiating power of the Sun
is known : a fair amount of it is received by
the earth; that is what is responsible for all the
activities on the earth,—the winds, the rain,
the rivers, the vegetation, and life generally.
But what the earth receives is but a minute
fraction of the whole : the Sun radiates in
every direction; the earth is only a small
body 92,000,000 miles away : the fraction
it receives is very little compared with the
whole. Even that is not insignificant however,—by no means insignificant to us : it
lies at the very root of our physical being,
though expressed as matter it is but small;
the earth receives from the sun about three
hundredweight a minute.

But the Sun is losing four million tons of matter every second! Accepting what we have said about the connexion between matter and radiation, the actual amount is a mere question of arithmetic. Arithmetic gives the figures—four million tons a second, lost to the Sun and radiated into space. Can the Sun stand such a loss without perceptible diminution? Yes, it makes no perceptible difference. The amount of matter even in the earth is 6000 trillion tons, that is 6000 million-million-million tons: the Sun has three-hundred-thousand times as much as that: and there is no difficulty in supposing that it has been losing, at much the same rate as now, for ten-thousand-million years. Less than one-tenth per cent. of its substance would have gone away in that time.

The antiquity of the Solar System is fearful. Life on the earth, in some form or other, has been going on most of that time: that is the conclusion to which we have come by studying electrons and the properties of the Ether, and digging down to the ultimate nature of matter. All that time, Life has existed; but not all that time Intelligence; not intelligent life as we know it now.

The Ages of the Earth's past seem to have been a sort of preparation for the life and mind which now is, and for the mind which is still to come. It has been a slow and laborious process; and the outcome of it, so far, we see. Has the outcome been worth all the labour and time spent in preparation? Faith is needed to suppose that. Mankind with its trivialities seems hardly worthy. Yet by faith we feel bound to suppose that there is some far-seeing design, some lofty meaning in it all, and that the ultimate outcome will be worth while.

Meanwhile here we are, conscious creatures endowed with the faculty of understanding; with the power of work, and of mutual help: we have become agents in the Universe endowed with consciousness. We are beginning to understand something of the processes at work : we are beginning even to have some power, not only of understanding but of helping, of doing things which would not otherwise be done. We have the power to produce works of art, to re-arrange matter into beautiful forms, to exhibit emotions, to glorify existence by human achievements, to express feelings and hopes and aspirations, to add to the real wealth or value of existence. And so, even while still within the range of the material universe, we may hope that mankind will become worthy of the magnificent scene in which its lot is cast. And we ourselves, though we are but a stage in the majestic pageant, can yet understand that here on earth is scope for free response to a Divine Will, and can wonder at the Patient Effort.

Depend upon it, nothing is easy, nothing petty, nor is anything haphazard, things are not left to chance. Everything is amenable to law and order, everything points to a rational Scheme or Plan, of which we know neither the beginning nor the end, but towards the fulfilment of which we can help. In face of all that, shall we allow ourselves to squabble about trivialities ! Shall we crawl about on the surface of the planet and sting each other here in the dust and die ? Or shall we realise that we are the heir of all the ages, that the destiny of mankind is being partly entrusted to us, and that humanity has a future, a potential future, beyond our wildest dreams !

Chapter X	Life and Mind and their use of the Ether

Chapter X   *Life and Mind and their use of the Ether*

The Ether's perfect properties, and service to Reality.

Is not animal motion performed by the vibration of this medium, excited in the brain by the power of the will?—NEWTON.

Every vital process may conceivably thus be correlated with a mechanical process . . . without lending any countenance to a theory that would place its initiation under the control of any such system of mechanical relations.—SIR JOSEPH LARMOR.

Whether this vast homogeneous expanse of isotropic matter is fitted not only to be a medium of physical interaction between distant bodies, and to fulfil other physical functions of which perhaps we have as yet no conception, but also . . . to constitute the material organism of beings exercising functions of life and mind as high or higher than ours are at present—is a question far transcending the limits of physical speculation.—CLERK MAXWELL.

LET us now run over some of the properties of the Ether so far as they have been ascertained. It is a universal connecting medium, filling all Space to the furthest limits, penetrating the interstices of the atoms without break in its continuity. So completely does it fill Space that it is sometimes identified with space; it has been spoken of as Absolute Space; it is also called "The Continuum." But, whatever name be given to it, it is a substantial reality; though to us it appears as empty space because we have no sense organ for its appreciation. It is invisible, intangible, inaudible. But it can be thrown into tremors by heated bodies, and for some of these tremors we have a receiving organ in the eye, which is our one physical or physiological link with the ether.

Matter has nothing to do with the transmission of light, it is an obstruction : it can check or modify the transmission, and that is all. Light is a purely etherial phenomenon. So also are electric and magnetic fields. They exist in connexion with matter, but in themselves are purely etheric.

The first function of the Ether is to weld the atoms together by cohesion, and the planets and stars together by gravitation. The second function is to transmit vibrations from one piece of matter to another,—which it does at a great but finite speed which can be measured; thus telling us something about itself, and showing that it belongs to the material or physical Universe, though it is not what we ordinarily speak of as matter. Matter is that which is capable of locomotion : its primary property is to move from place to place; though the inertia or momentum with which it moves must belong wholly to the Ether of which it is presumably composed. People sometimes try to feel a difficulty about motion in a plenum; they say, If space is completely full, how can things move? There is no real difficulty : fish move about freely in the depths of the ocean.

The Ether is now believed to be a very substantial substance, far more substantial than any form of matter. Matter is made of separate particles, at a considerable distance from each other : it strikes us as a sort of accident in the Ether, with a structure which demands explanation and investigation, not yet completely forthcoming.

The Ether has perfect properties, such as no form of ordinary matter has. Matter is only sub-permanent; it is liable to all manner of deterioration. Solid bodies break up, disintegrate, and decay : material things wear out, they are imperfectly elastic : a spring bent too long gets permanently set, and does not recoil. Matter is subject to fatigue; it ages: complicated molecules break up into simpler ones. Nothing in that sense is permanent. The everlasting hills are not everlasting; they rise and they fall: the changes are slow but inevitable.

> " The hills are shadows and they flow
> From form to form and nothing stands.
> They melt like mist, the solid lands,
> Like clouds they shape themselves and go."

The crust of the earth displays the history of past times.

> " Oh, Earth, what changes hast thou seen!
> There where the long street roars, hath been
> The stillness of the central sea."
>
> [In Mem. 123.]

Of ancient civilisations we find only traces: the most solid buildings are temporary. All the energy of matter tends to fritter tself down into heat,—what is called the dissipation of energy. Not that the energy goes out of existence, it changes its form and ceases to be available; like milk spilt upon the ground, it is no longer useful, though still it exists.

It used to be thought that this law of dissipation applied to the whole Universe: it applies only to the matter portion of it.

The stars are wearing themselves away; where there is heat there is also radiation; the energy of matter is transferring itself to the Ether, which is the universal store-house of energy. Whether matter is ever reconstituted from the Ether we do not know: if it is, it will be re-made in the depths of space, far away, and can then fall together again and renew another and another frame of things for ever.

Energy is constant in quantity; it changes its form. Sometimes it is matter in motion, then again it is ether in vibration, or else it is strained ether without vibration. All the energy really belongs to the ether but it manifests itself in different forms. No law of dissipation applies to the Ether; that is what I mean by saying that its properties are perfect. Ether fritters away no energy, is preserves all: it is perfectly transparent; it transmits light from the most distant stars without waste or loss of any kind. Only matter dissipates energy, only matter wastes and decays, only matter ages and wears out.

What we know as electric charge and magnetism can readily pass into insensible forms. A steel magnet may lose its mag-netism, just as a steel spring may lose its elasticity; a charged body can become discharged; but the essential ingredients remain. The magnetic loops, which in a magnet are opened out and made perceptible, never cease; they may close up, but they are still there, and the circulation with which they are endowed continues without friction and without loss. So also the ripples or tremors in the Ether, however much they may become diluted by spreading out, and however feeble they are by reason of distance, never cease : they either continue as ripples, or they turn into some other form of energy. They are, astonishingly, still found to have the power of ejecting electrons with full vigour, when they encounter atoms, even very distant atoms, under proper conditions.

It is important to recognise that the Ether and its properties are absolutely permanent : there is no irregularity or random motion in ether, as there is among the atoms of a gas or solid. Random motion in matter is called " heat " : there is no true heat in the Ether : there is vibration, but it is systematic and orderly vibration, travelling always at the same speed, the one absolute speed in the Universe, which we have measured as the velocity of light.

So far we are on fairly safe ground. Now let us speculate a little, and apply what we have learnt to the elucidation of some phenomena which we have not yet dealt with.

Matter exists not only in the inorganic form of solids, liquids, and gases, and in the disintegrated form of electrons and protons; it exists also as the complex molecules known as *protoplasm*, which for some reason or other has shown itself to be the vehicle of life. Some forms of matter are endowed with or animated by life. This property of animation is a great mystery; we do not know what " life " is, we only see what it can do. We perceive that it can enter into relation with matter, that it has a character and identity of its own, and that it builds up matter to correspond with or to represent that identity. Life can take a variety of forms, and every form is characterised by a certain shape : the life of an oak is transmitted to an oak, the life of an elm to an elm. " To every seed his own body." One form of life takes the shape of a bird, another of a fish, another of a quadruped. The varieties of life are innumerable, and are studied in the great science of Biology.

What has that to do with the Ether?

Consider any piece of matter and we shall see. Contemplate any solid object; a vase, it may be, or a jewel, or a statue : what is it that holds the atoms together in that particular shape? If the atoms were not connected they would be moving about at random, like the atoms of a gas; but they are connected, crystallized as it were together, by the forces of cohesion. Even in a liquid they are held together into a body of definite size, though not of a definite shape. A liquid has size, though not shape; a gas has neither; a solid has both. The shape is most definite and law-abiding in a crystal; but in a plant or an animal it has a definite character too,— not so definite as in a crystal, a good deal of variety is possible, yet an animal or vegetable body has an undoubted character of its own, even to minute detail. And this character is handed down from one generation to another; modified perhaps, but only slowly, by the age-long process of Evolution.

At a certain stage in the course of evolution, not Life only makes use of the animated protoplasmic material, conscious Mind enters into relation with it too. There is much to be said about that; nor is the subject free from controversy. But however the fact ought to be expressed, the fact itself is familiar. Life and Mind have entered into relation with matter. What they are we know not : we can only study their behaviour : they use matter for a time and then disappear. " Disappear " : that is the word : we have no right to say they go out of existence; that would be going beyond our knowledge : they go out of our ken. If they are real things, it is quite unlikely that they go out of existence. What their existence really means I for one have no idea : all I know is that they can temporarily animate and control matter.

But do they animate matter alone? What about the Ether which holds the atoms together, the welding ether which is essential to the characteristic configuration of a body —which is as essential as the matter itself? We do not usually attend to the ether aspect of a body; we have no sense organ for its appreciation, we only directly apprehend matter. Matter we apprehend early, when young children, but as we grow up we infer the Ether too, or some of us do. We know that a body of characteristic shape, or indeed of any definite shape, cannot exist without the forces of cohesion,—cannot exist therefore without the Ether;—meaning by

the Ether now, not the whole, but the un-
materialised part of it, the part which is the
region of strain, the receptacle of potential
energy, the substance in which the atoms of
matter are embedded. Not only is there a
matter body, there is also an ether body :
the two are co-existent.

Does Life enter into relation with the
Ether as well as with matter? How it
relates itself with matter we do not know;
we only know the fact. We cannot assert
that it enters into relation with Ether : we
can but ask the question. A matter body is
animated when it belongs to a plant or an
animal. Is the Ether body likewise animated? If so, we may well ask further
—what happens when the matter body wears
out? We may be sure that the Ether body
does not wear out : that is contrary to all
we know about the Ether and its perfect
properties. No, but if the animation ceases
from one, it may cease from the other. It
may : we cannot assert either way : it is a
question of fact; and the fact is not yet
certainly known.

But there is a higher kind of animation,
that which is characterized by Mind and
Consciousness, and Memory and Affection,

162

and many other extraordinary attributes, which we know of but do not understand. I do not deny these attributes to the higher animals, but they are conspicuous in man. Are they transitory, or are they permanent? In the first instance we only ask the question, and say it is a question of fact still to be ascertained. These attributes do not seem to belong to the material or physical universe at all : they seem to belong to another order of things—(perhaps " things " is not the right word)—another order of existence : we do not understand their nature, but up to a certain point we are familiar with them. They belong to the Unseen Universe, the universe which makes no appeal to our senses. Do these psychical attributes require a vehicle in the material universe? Apparently they do : we know them when embodied, they are embodied in matter; they act on matter and move it, move it and rearrange it,—which is a special kind of motion. That is all they do to matter : they make use of the particles of matter to display themselves, and only when they operate on matter are we aware of mind and consciousness in others, because our senses are material senses.

163

But, we are bound to ask, about these psychical things, do they act on matter directly or indirectly? Now we are coming to a region which is open to experiment and observation. How do we ourselves act on matter? The matter of our bodies has been put together unconsciously : our individuality has done it, in ways we know not how : we could not build up this body consciously. But it has been built up, and it is ours to control. By aid of the body we can operate on external matter : we can consciously move other bodies; we can build things, we can set things rolling, we can lift things, we can also imagine structures and then make them; we can design and we can execute; we can build bridges and cathedrals; we can paint pictures and carve statues; we can throw matter into vibration and produce audible music; we can move a pen over paper and write a poem; we can speak and we can broadcast. In other words we can use matter to display our thoughts. Moreover beings similar to ourselves can receive those thoughts, being acted upon by the vibrations of the air or of the Ether : they have the faculty of interpreting those vibrations in the way which the originating mind

164

intended. All this is very puzzling, very mysterious. The Universe to which we belong is greater than the physical universe : it utilises and dominates it. Shall we say that we share the disabilities of matter, that we wear out and decay, that our existence is limited to the instrument which we employ ? Or may we surmise that, like the Ether, we have a more permanent and perfect existence, not liable to death and decay ? It is a question that we can ask, to which an answer may in due time be forthcoming : in my opinion has already come.

Meanwhile I have not answered the question which I asked a little time ago. How do we consciously, through our muscles, act upon matter ? The first answer is, By contact : we touch a body and we move it. But consider what we have learnt. What do we mean by contact ? Atoms of matter are never in contact : when two pieces of matter come within close range of each other, there are forces of repulsion between them which prevent contact : one electron cannot touch another electron ; they repel each other too violently for that. Whether an electron can ever touch a proton we do not know, but if it did, something extraordinary would happen ;

there would be a flash of radiation and the two particles would disappear. That is not what happens when we move a body! The fact is *we touch it only through the Ether*. Just as a magnet attracts a bit of iron through the ether, and an electric charge repels another through the ether, so it is on the ether that our muscles act directly, and on objects only indirectly. I believe that that is so always, and that our real bodily manifestation is through the Ether primarily, and through the matter associated with it indirectly. I wish to make the hypothesis that it is the Ether which is really animated, and that this animated ether interacts with matter; I suggest that the true vehicle of life and mind is Ether, and not matter at all.

But so long as we acted on the ether only, —if our action was limited to that—we should not be able to make any impression on the senses of our friends and neighbours : hence indirectly, and by means of the ether, we act on matter as well. Somehow or other we have constructed bodies which represent our personality; and with them we can move about, make vibrations, alter the configuration of the world in which we live, and represent to others our ideas and conceptions and feelings and wishes and desires. It is a very indirect and singular process : there may be other and more direct methods of mental communion : some people think there are, and call them Telepathy; others think we can only act through and by means of matter. Somehow or other one mind can act on another, but the ordinary method by which it does so is extremely roundabout : in some mysterious way it liberates energy from the brain-cell, which then travels along a nerve, stimulates a muscle to contract; and then either the hand writes, or the larynx vibrates, or the fingers press a telegraph key. The result is that either an aerial or an etherial disturbance is set up, which travels to a distance, is there received by a suitable instrument (usually either an ear or an eye,) appropriate nerves are stimulated, and the stimulus reaches a brain cell. Energy put in at one end comes out at another, and in some quite unknown way a corresponding thought is aroused in the receiving mind.

There is delay in the process; time is taken for the vibration to travel—the thought seems to have to exist in a curious mechanical form in its passage between the organisms— but the delay is usually not great. There

167

may however be great delay; the material impression produced by one person may be delayed in reception till long after he is dead. Someone has made marks on a bit of paper, or has arranged pigments on a canvas; and this rearrangement of matter can be buried for a century, to be interpreted only by a subsequent generation. Not to be interpreted at all unless submitted to a competent mind.

These are the strange phenomena to which we have grown accustomed. Certainly the *records* can survive the death of the body; but can the thought which produced them survive, can the conscious memory of events survive? I say that if the Ether is animated, and if the mind is acting on the Ether and uses it as the real vehicle, they have every chance of surviving. Mind may always need a vehicle, a body, a habitation; but it need not be made of matter. If what I have suggested is true, it is not really made of matter now: it belongs to the insensible world, the world beyond our senses. Only for purposes of transmission and communication do we need the world of matter: our real existence is elsewhere and otherwise. We—our own nature—must not be confused with the atoms; they are an intermediate tool, a weapon, an instrument, a means of manifestation. Music conceived in the mind can be reproduced on an organ or a violin, but music itself is not dependent on those instruments; they are only used for the purpose of making it audible to others. The Ether is a permanent vehicle, probably adapted to the utilisation of something still more beyond our senses than itself. Reality lies in the unseen, the permanent; where there need be no imperfection, no wearing out and decaying, no dissipation of energy, no loss or waste or fatigue. All these imperfections belong to the assemblage of atoms which we call matter: these truly are temporary, but Reality is permanent. " The things seen are temporal, unseen things are eternal."

Y*

Spiritus intus alit, totamque infusa per artus
Mens agitat molem, et magno se corpore miscet.
                                                    VIRGIL.

THE Ether of Space is a theme of unknown and apparently infinite magnitude,
and of a reality beyond the present conception of man.   It is that of which every-
day material consists, a link between the
worlds, a consummate substance of over-
powering grandeur.   By a kind of instinct,
one feels it to be the home of spiritual
existence, the realm of the awe-inspiring
and the supernal.   It is co-extensive with
the physical universe, and is absent from no
part of space.   Beyond the furthest star it
extends; in the heart of the atom it has
its being.   It permeates and controls and
dominates all.   It eludes the human senses,
and can only be envisaged by the powers of
the mind.

Yet the Ether is a physical thing, it is
not a psychical entity, it has definite physical

173

properties. It is not matter, any more than hydrogen and oxygen are water, but it is the vehicle of both matter and spirit. Its mechanism is unknown to us, its inner nature eludes us; yet mechanism it must have, for it is subject to physical laws. Its vibrations can be analysed : they bring to us information, and without them we could not exist. All life on the planets is dependent upon Ether tremors; Matter is responsive to them, and they may have been instrumental in bringing matter into existence. Ether is the universal connecting link; the transmitter of every kind of force. Action at a distance is wholly dependent on the Ether, and it is manifestly the vehicle or substratum underlying electricity and magnetism and light and gravitation and cohesion.

But though it is the seat of all electric forces, and indeed the sole transmitter of force, ether is not electricity any more than it is matter. Yet an electric charge must be composed of it, in a way which has yet to be discovered. Undoubtedly it has electrical and magnetic properties, and is the vehicle of gravitation and of light. It welds the planets into a solar system. It unites the parts of an atom, and it holds the atoms

174

together. It is the seat of prodigious energies—energies beyond anything as yet accessible to man. All we know of energy is but the faint trace or shadow or overflow of its mighty being.

Hidden away in its constitution is a fundamental and absolute speed, a speed not of locomotion but of internal circulation. What it is that is thus whirling we do not know : without the whirl we have no conception of it. The whirl and the fundamental something together make up the Ether. And we have no power of detaching the one from the other, hardly even in thought.

Over great realms of space there is nothing else; but here and there is a modified portion, so modified as to be the seat and subject of what we call gravity, but exactly how modified we have still to find out. We know the unit of this portion as a localised singularity, capable of position and of locomotion. Can it be a special kind of whirl, or is it a knot or a strain or a bubble, a hollow or an extra condensation, or what? This is a question to be tackled; but as yet we cannot fully answer. In one form we call such individualised unit an electron; and we know that of such units, and of units akin

to them, the atoms of matter are made. We
know that all the bodies we see and handle
are but elaborate and beautifully organised
congeries of positive and negative electrons,
held together and connected by the medium
of which they themselves consist. The world,
the stars, the heavens, are nothing else. The
mystery of existence is close upon us here.
We cannot claim knowledge. We grope in
a kind of helplessness with our few animal
senses, and we live our short animal lives,
encouraged by a faith and hope that we are
something more than appears, and that in the
deep roots of our being we belong to another
order of things, which is associated with this
physical order perhaps only for a time.

Speculatively and intuitively we feel to
be more in direct touch with the ether than
with matter. How we can act on matter is
a mystery. How we have constructed and
how we move our bodies, we do not know.
We are apt to identify ourselves with our
bodies. But there is evidence which shows
that we are really independent, that we continue in existence, and can leave our bodies
behind.   Matter is not part of our real being,
not of our essential nature : it is but an
instrument that we use for a time and then

discard. Probably we do not act directly upon matter at all. Our will, our mind, our psychic life, probably act directly upon the Ether; and only through it, indirectly, on Matter. Ether is our real primary and permanent instrument. It is in connexion with the Ether that our real being consists; and through it we are able to manipulate the atoms of matter, to move them, to rearrange them, and thus employ them to express our thoughts and feelings and to manifest ourselves to other individual entities who in the long course of evolution have been enabled to construct and employ similar most ingenious, though imperfect, instruments of manifestation. By this means we can become aware of a multitude of existences, the whole animal and vegetable kingdom, of which otherwise we might have remained ignorant; by this means our conceptions of existence have been enlarged and extended, the possibilities of friendship enhanced, the perception of a new realm of law and order attained. And thus is our own nature enriched by the effort and experiences belonging to a new and most interesting—though from our point of view imperfect and rebellious—physical mode of existence.

While thus incorporated in matter we communicate with each other through signs and codes, through symbols and material means. We send messages by vibrations of the air; we record them by conventional traces. We preserve and study the marks— the writings, records—made by previous generations : we utilise for signalling purposes, not only the vibrations of the air, but the vibrations of the Ether. The animals have long learnt to do this through their astonishing sense organ the eye; which indeed appears very low down in the scale of evolution, and may be regarded as an outstanding sign of our etherial connexion. The ear appears as a later development : we learned to employ aerial vibrations later. The Ether connexion is simpler, more direct, more informing, less dependent upon code, more immediately intelligible than anything connected with language. Pictures appeal to children before words. Pictures made appeal to very early humanity before language. Visible things were apprehended before pictures. Thus it is through the Ether that we get our earliest information.

Touch seems to be a purely material sensation, the result of direct contact with

matter : it is indeed what we call " contact."
But when we come to analyse touch, we
learn that atoms are never in contact.
They approach each other within an infinitesimal distance; but there is always a
cushion, what may be called a repulsive
force, between them,—a cushion of Ether.
Hence even our apparently most material
sense is dependent on this omnipresent
medium, on which alone we can directly act,
and through which all our information comes.
It is the primary instrument of Mind, the
vehicle of Soul, the habitation of Spirit.
Truly it may be called the living garment of
God.